WITHDRAWN

THE
ELECTRIC
WISHING
WELL

THE ELECTRIC WISHING WELL

THE SOLUTION TO THE ENERGY CRISIS

Joseph J. DiCerto

Drawings by Joseph J. DiCerto

MACMILLAN PUBLISHING CO., INC.
New York
COLLIER MACMILLAN PUBLISHERS
London

Copyright © 1976 by Joseph J. DiCerto

All rights reserved. No part of this book may be reproduced or transmitted in any form or by any means, electronic or mechanical, including photocopying, recording or by any information storage and retrieval system, without permission in writing from the Publisher.

Macmillan Publishing Co., Inc.
866 Third Avenue, New York, N.Y. 10022
Collier Macmillan Canada, Ltd.

Library of Congress Cataloging in Publication Data

DiCerto, J J
  The electric wishing well.

  Bibliography: p.
  Includes index.
  1. Power resources. I. Title.
TJ163.2.D5   333.7   76–18885
ISBN 0–02–531320–7

FIRST PRINTING 1976

Printed in the United States of America

To Nina,
truly my better half
and to Lisa, David and Jennifer—
three very understanding children

Contents

Preface

Although there is an energy crisis, there is no lack of energy. If there were, the sun would not shine, the winds would not blow, and the seas would not rise and hurl their waters in waves and tides. The earth's core would be stone cold and the seas would be frozen over. But the center of the earth is hot with life, the oceans and the air teem with movement and the sun shines steadily. There is a magnificent display of abundant energy in, on, and around the earth.

The energy crisis is actually a people crisis. The individuals and societies of the earth have managed, through thoughtlessness, unconcern and greed, to trap themselves in an untenable position. They have established their industries and economy on fossil fuels which, it turns out, are critically limited in supply, both geographically and quantitatively. A little foresight and fundamental mathematical extrapolation would have discovered this fact many years ago. Now, we are rapidly running out of the fuels needed to generate the energy crucial to a modern industrial society. But it appears we are slow to learn from experience.

Some would have us "solve" the energy crisis through the extensive use of nuclear fission power plants, which are presented as the only alternative capable of meeting the rapidly increasing energy demands of the coming decades. However, with more foresight, closer inspection and a little mathematical extrapolation, the long-range effects of widespread nuclear power can be envisioned and the picture is bleak indeed. Massive use of nuclear power could plunge mankind into a series of crises of devastating magnitude.

That we inhabit a planet which offers a virtually inexhaustible supply of energy—far more than we could ever use—is the message of *The Electric Wishing Well*. Huge quantities of clean, reliable energy are available. We are standing before a mythical wishing well, asking for energy—the basic substance of life—and that wishing well is answering us. Its message is transmitted to us, not in words, but in the natural media of sunbeams, wind, heat and other forces. For those who do not understand the language, this book provides a translation; for those who already understand the language of the Electric Wishing Well, this book offers a challenge. How seriously do we view the present and future energy situation? Are we placing monetary gain above the welfare of society? Are we willing to work together on a national scale to make the transition from a fragmented, disorganized and destructive energy approach to a well-coordinated and integrated system that offers clean energy in abundance?

JOSEPH J. DICERTO

THE
ELECTRIC
WISHING
WELL

1

The Electric Wishing Well

It's frightening, but it's true. Right now we are depleting the earth's fossil fuel supplies about a million times faster than nature can replenish them. Nature cannot keep pace with man's ever-growing appetite for energy. Man seeks more and more and at times reaches for what is beyond the practical limits of his capability. What he cannot reach, he wishes for. He stands hopefully at the proverbial wishing well, an energy wishing well, as he has done for a hundred millennia, trying to conjure up new miracles.

Even Cro-Magnon man was acutely aware of the many benefits he derived from the natural world around him—the sun, for instance, that colossal light far above his head. Because of its light, he could see the beautiful and frightening world around him; he was comforted by the warmth of its heat on his body. Yet he was not fully satisfied with these gifts. All too soon the day lost its brilliance and with the blackness of night came fears of the unknown, terrible sounds from nocturnal predators and the numbing chill of a damp cave.

Early man was no stranger to the wishing well. "I wish I could have the power of light and warmth. I wish I could reach up and grasp some of the heat and light from the sky-bound power that flees with the coming of the dark." Nature took pity on man and his wishes were granted.

It is unknown to us how man first domesticated fire. Perhaps it was the result of lightning; maybe the cause was volcanic activity. Nevertheless, early humans discovered that a hard material, the same material they used for clubs, could be made to burst into a bright light that issued comforting heat. No longer was the cave a nightly black chamber where beings silently groped about, awaiting the return of the sun. With fire, man could operate on his own schedule; cave dwellers could mend their tools at night, sit together and socialize, while enjoying physical comfort and a sense of independence of the day-night cycle of nature.

Wood was the first of nature's fuels. Surrounded as he was by seemingly endless tracts of forest, it appeared to man that there was no limit to the fuel supply. By the time of the Babylonian, Greek and Roman civilizations, man had also learned to produce light from various types of fats and vegetable oils, and from pitch, a black oily substance from the earth.

Lost in the mist of unrecorded history is man's first encounter with a black rocklike substance that could burn with a very hot flame. There is archaeological evidence that coal was used by inhabitants of Glamorgan, Wales, during the Bronze Age, 3000 to 4000 years ago, and that Roman soldiers in Britain burned coal to heat their quarters as early as the fourth century A.D. This fossil material provided a rich source of energy, but one that would not be used extensively until the seventeenth century. The energy wishing well was beginning to give man a wider variety of choices.

As far back as 600 B.C. the first clues that signaled a significant change in the nature of this wishing well could be observed. The Greek philosopher Thales of Miletus amused friends by performing an amazing feat of magic: using an amber rod, he made straws leap from a table to the rod without touching them. Today we know this force as static electricity. Perhaps Thales thought about putting this force to practical use. But that would not happen for over 2000 years.

In 1600 an English physicist named William Gilbert published the first descriptive treatise on the theory of electricity, *De Magnete*. It earned him the renowned title Father of Electricity. During the next 150 years scientists performed various experiments to further identify the elusive force of electricity, but there was not much progress in its practical application. In the middle of the eighteenth century wood was still the major source of fuel, though other sources were beginning to appear on the scene. Because of his dissatisfaction with wood, man turned his attention to nature's more advanced forms of fuels—the fossil materials oil and coal.

In 1556 the German mineralogist Agricola published a treatise describing the extracting and refining of petroleum. In the late Middle Ages the first mine shafts had been dug to extract coal from beneath the earth. Progress in coal mining was slow and most operations were quite small. Records show that in 1684 the area around Bristol boasted 70 mines, but only 123 workers were employed in all these mines. Undaunted by the slow pace of coal mining, man continued his search for new ways to exploit the benefits of fossil fuels.

The invention and development of steam engines in the early eighteenth century accelerated and vitalized the industrial revolution in Europe. Now large amounts of energy were needed not only to provide warmth and light, but also to run the huge machines that labored night and day making products to be sold in an ever-expanding marketplace. The stage was set for the steady climb of energy consumption to extraordinary heights. From their simple beginnings as sources of heat and light, fossil fuels soon became the raw material for a hundred industries, from gasoline, lubricants and paints to cosmetics, chemicals and plastics. Today the U.S. economy runs on fossil fuels. We use them by the millions of tons, billions of barrels and trillions of cubic feet. And as huge as these numbers are now, they will continue to grow as the nation's appetite for energy swells to ever greater proportions.

THE LION'S SHARE

Over the past 100 years U.S. energy consumption has been increasing at a rate of 3 percent per year. This may not sound like a big increase, but by 1972 Americans were using energy at the staggering rate of over 71 quadrillion BTUs (British thermal units) per year (that's 71,000,000,-000,000,000 BTUs, or in more familiar terms, 71 million billion BTUs. Our per capita consumption is the highest in the world. It is twice as much as Great Britain's, two and a half times as much as Germany's, and four times more than Japan's. We use about one-third of all the energy generated in the world.

In certain parts of our country energy use reaches astronomical levels. Manhattan is a prime example. On a hot summer day the sun floods the city with a great deal of energy—about 93 watts per square meter (approximately the power consumed by a 100-watt bulb). Yet the heat released from Manhattan's energy consumption averages 630 watts per square meter, more than six times that derived from the sun.

If we project the trend in energy consumption to the end of the century, we'll discover that we will be burning up 192 quadrillion BTUs per year. The following example will put this number into perspective. 192 quadrillion BTUs is the equivalent of about 40 hp (horsepower) for an eight-hour day for every man, woman and child in the United States. A man hard at work produces a maximum of $\frac{1}{20}$ hp in eight hours; the per capita availability of 40 hp is like having 800 energy slaves. Where will all this energy come from in the next few decades? To quote Russell W. Peterson, chairman of the President's Council on Environmental Quality, "That amount of energy production is simply out of the question without unacceptable dependence on foreign sources or unacceptable damage to our environment."

If the energy were derived from coal, we would have to dig up half the United States. If from oil, we would have to drill enough holes to make Alaska, the Southwest and other areas of the country look like Swiss cheese. Gas alone could not do the job. The United States would have to import a great deal of oil—about 25 million barrels per day—and liquid natural gas. We would also have to operate about 1000 huge atomic power plants with all their attendant problems.

One response to the energy crisis is to do more of what we are presently doing—that is, drill more oil wells, pump more natural gas out of the ground, mine more coal, build more atomic plants and import more fuel. By the end of this decade fossil fuels will still be carrying the preponderant burden of the energy load. Atomic power will supply about 8.5 percent of our energy needs. Except for the insignificant contribution of geothermal power (less than 0.5 percent), none of the unconventional energy sources, such as wind, tidal and solar power, will be exploited at all. Although oil, gas and coal are valuable as raw materials to the chemical, plastics and synthetic fibers industries, a significant amount of these materials will continue to be burned to produce heat to generate electricity. That our society is firmly committed to its present energy-producing system is well illustrated by the following tables.

In 1972 the United States consumed energy from fossil fuels in these proportions:

Oil	45.5%
Natural gas	32.3%
Coal	17.2%
Hydropower	4.1%
Nuclear power	0.9%

Oil, gas and coal carried practically the entire energy load (95 percent). Atomic power contributed less than 1 percent. And a quarter of a century from now (in 2000), according to a projection by the U.S. Bureau of Mines, these fuels will still account for the major portion of our energy (69.6 percent). The fuel mix is projected as follows:

Oil	35.1%
Natural gas	17.2%
Coal	16.3%
Nuclear power	25.4%
Hydropower	6.0%

Such a strong commitment to fossil fuels has far-reaching ramifications in terms of depletion of resources, political consequences, environmental effects and economic results. We are quickly running out of easily accessible fossil fuel supplies. We are becoming increasingly dependent on foreign countries for critically needed fuel. We are now in strong competition with the European countries and Japan for fuel. We are rapidly degrading our environment (air and water pollution, land destruction). And the cost of obtaining energy is rising sharply, thereby affecting our ability to sustain a strong economy.

The logical solution to these problems is to evolve to a totally electric society and a hydrogen economy. This proposal may appear startling and fanciful. In order to understand and have a basis for accepting it, we should take a closer look at the various sources of energy in the United States today.

OIL

As far back as 3000 B.C. the Sumerians and the Babylonians were using a black substance that oozed from the earth. It came to be known as "petroleum" from the Greek and Latin words for "rock" (*petra*) and "oil" (*oleum*). Originally, it was used to seal or caulk ships to make them watertight. Almost 5000 years passed before man had the technological skills to pierce the earth and reach far below ground for this precious fluid. Once again, man had approached the wishing well with a desire for additional energy, this time in large amounts. He had long ago mastered the use of fire for light, heat, preparing food and producing products, he had burned vast amounts of wood and he had discovered the magic of burning coal. But oil held out the promise of even vaster amounts of energy. And the great industrial revolution produced thousands of energy-consuming machines to be run and lubricated, millions of lamps to be fed and furnaces to be fired.

Driven by these forces, a pioneer of the petroleum industry, E. L. Drake, drilled the first oil well in western Pennsylvania in August 1859—25 years before the development of the automobile. The depth of this historic well was only 69½ feet, a modest beginning for what was to become a vast industry and a lifeline of our economy. By the end of the nineteenth century oil was being used for a strange new contraption that moved without the assistance of a horse. The gasoline that propelled this machine along the road came from refined oil. As more cars took to the road, the demand for oil began to rise, and it has not stopped rising since.

How fast and how great has been the rise? By 1950 the citizens of this planet were using about 11 million barrels each day—7 days a week, 365 days a year. That means that we were sucking over 4 billion barrels of oil from nature's storehouse in one year.

The United States in 1950 was using over 6½ million barrels of oil each day, more than half of the world's consumption. And we kept asking for more. In 1955 we jumped to 8.4 million barrels each day; in 1965 to over 11 million barrels per day. By 1970 nearly 19 billion barrels of oil were pumped out of the earth each year, and the United States was using over 6 billion of them. In 1980 the inhabitants of this world will be using an unbelievable 29 billion barrels of oil each year. During the 110 years between the drilling of the first oil well and 1970, approximately 225 billion barrels of crude oil had been extracted by the world petroleum industry. But in only one decade, 1970–1980, the world will consume almost the same amount that it did during the first 110 years of the petroleum industry.

Where is all this oil going? Much of it is consumed in transportation. The 90 million cars, trucks and buses speeding along our highways, the speedboats and cabin cruisers, and the private and commercial aircraft burned up 93 billion gallons of gasoline in 1970. By 1974 the figure had soared to well over 100 billion gallons. Then there is the heating of homes and stores (including hot water). In 1973 that accounted for over 1 billion barrels of oil a year.

Oddly enough, a major factor in the increased demand for oil was pollution. Environmental constraints have important effects on our use and choice of fuels. For example, automobiles have been a major cause of air pollution. When federal laws were passed to limit the amount of pollutants that a car could emit, antipollution devices had to be added to car engines, and as a result, engine efficiency and gas mileage dropped. So, cars burned more fuel and the demands on the petroleum industry rose. Antipollution laws also caused many factories to switch from coal, which generated a great deal of pollution, to oil, which burns much more cleanly.

One of the biggest factors in the increasing demand for oil has been the transition from burning coal to burning oil by electrical utilities. Today about 40 percent of all oil used in the U.S. is consumed by the electrical power industry. A good example of this shift can be found in the New York metropolitan area. As late as 1960 only 22 percent of the fuel used by Consolidated Edison and Long Island Lighting was in the form of oil. By 1971 that figure had jumped to 80 percent.

The switch to oil has also taken place in Europe, which has historically relied on coal as its major fuel. Before 1940 coal supplied about 80 percent of Europe's energy. Thirty years later that figure had dropped to 28 percent, the balance being supplied by oil and natural gas.

For the United States, which is one of the world's major producers of oil, the picture is not encouraging. At the rate we are consuming this resource, we have only about a seven-year supply of *easily* accessible oil remaining to us. The projected 90 billion barrels of oil from Alaska's

North Slope represents only about a three-year supply of oil at the increased rate of consumption. If we cannot learn to do without oil, our only recourse in the near future is to import vast amounts from Middle Eastern and South American countries. But this will be extremely costly, since the oil-producing nations have raised the price of crude oil so dramatically. In 1972 a barrel of crude oil sold on the open market for about $3.39. In October 1973 the posted price of oil on the world market was actually down to $3.01 per barrel. Three months later, in January 1974, as a result of the Arab boycott, the price skyrocketed to $11.65 per barrel, and many countries' bills for imported petroleum more than tripled in one year.

What makes this "black gold" so valuable is that over the past few decades we have practically built our economy on it. Not only is almost all transportation dependent on oil, but almost half of our electrical generating capacity depends on it as well. In addition, oil is crucial to a host of industries. In agriculture for example, petroleum is crucial not only as the energy used on farms and in the processing of farm products, but also as a raw material for fertilizer, pesticides, insecticides and herbicides. The food production capabilities of the United States and the world depend greatly on the availability of large amounts of fertilizers. One of the results of the energy crisis is a scarcity of American fertilizer for export to the undeveloped nations. The shortage of oil translates into insufficient food for the poor nations. Kenneth E. Baulding, professor of economics at the University of Colorado, has asserted that "Oil indeed may be more important in the future as a source of fertilizer than it is of gasoline, and we are already beginning to see some indications that this [shortage of fertilizer] may be the most drastic impact of an oil shortage."

Each year over 400 million barrels of oil are used for nonenergy applications—that is, processed for the production of products. Perhaps most dramatic, in terms of numbers, is the relation of oil to the plastics industry. This industry is highly dependent on petroleum as a raw material. By 1985 worldwide production of plastics will be 273 million tons, and by 2000 it will swell to 1.7 billion tons. In the United States alone 40 billion pounds of plastic will be produced each year during the 1980s. This plastic will find its way into millions of different products in our homes, offices and cars, and on our persons. The average car will contain about 200 pounds of plastic. Fifteen billion molded plastic bottles will be produced each year. Clothes, tools, toys, furniture—the list is endless. And this entire industry is absolutely dependent on petroleum and petroleum-related products.

So petroleum is obviously a very precious commodity, certainly not one to be used for so basic a purpose as generating heat. We burn it in cars, in hot-water boilers, in furnaces to heat our homes—and these jobs could be done by other forms of energy.

The big question is, will we have enough oil to meet future U.S.

needs? The answer depends on how much new oil we find, and how we use the oil we have. It is estimated that half of our domestic oil reserves are yet to be discovered. According to the National Petroleum Council, the undiscovered oil reserve is approximately 810 billion barrels (enough for about 100 years). This includes all the oil that will be found in Alaska and off the U.S. coastline. Bent on finding and extracting this huge amount of petroleum, oil companies are spending hundreds of millions of dollars on exploration and new equipment. Today about 15,000 geologists are searching for oil, using the latest pieces of electronic equipment, 5000 geophysicists are interpreting the findings of the geophysical surveys and more than 20,000 engineers are applying their technology to extract the oil.

In addition to the 810 billion barrels we may expect to find in the ground, there is another form of fossil material that may extend our oil supply for decades: shale. Shale is a soft stone impregnated with oil. It is found mainly in the Rocky Mountain states of Colorado, Utah and Wyoming. The idea of squeezing oil out of rocks is promising. First, the supply is right here in the United States and thus is not subject to foreign control. Second, we know exactly where it is located, so a lot of money and time need not be spent in exploration and drilling. Third, shale oil has a very low sulphur content and therefore is considered one of the cleaner fuels. And there is a lot of it. Estimates of the oil content of the western shale deposits range from about 54 billion barrels to trillions of barrels. How much oil can we squeeze or, more accurately, boil out of these rocks?

There are different grades of shale. The best contains as much as 40 gallons of crude oil per ton. The poorer grades offer as little as 15 gallons per ton. Until recently, the idea of extracting oil from shale was not very attractive because the process cost about $6 per barrel, while crude oil was selling for about $3 per barrel on the world market. But once the price of oil shot up to $10 per barrel, shale oil began to look more interesting. And recently the cost of processing shale has dropped to less than $5. That makes it a star candidate for the oil-producing companies.

The process of extraction is basically simple. The shale is mined from the earth. Large chunks are then crushed and placed into a special tank called a retort. In the retort the shale is heated to about 950° F., causing its oil to boil off. The oil vapors are then cooled and collected for further refining. Another method of extracting oil from shale is called in situ— "underground." Natural gas is pumped into a cavern deep below the earth in the middle of a shale deposit. Then the gas is ignited. The inferno heats the surrounding areas, causing the oil to be released. A well is then drilled into the shale deposit and the oil is pumped out.

Unfortunately, progress is slow and no large processing plants have been built yet. The National Petroleum Council estimates that only

750,000 barrels of shale oil will be produced per day by 1985. That will probably fill about 3 percent of our needs. More optimistic is Dr. Armand Hammer of the Occidental Petroleum Corporation. He believes that large quantities of low-cost, low-sulphur oil can be produced from Rocky Mountain shale before 1980, and that as much as 2.9 million barrels per day, or 11 percent of the nation's projected oil consumption in 1985, could be produced from these deposits. This would certainly help the United States attain its objective of self-sufficiency in energy.

But extracting oil from shale is not without problems. First of all, a great deal of water is required for processing. Unfortunately, there is not an overabundance of water in the shale area (Washakie Basin, Green River Basin, Uinta Basin), and this lack could limit the production of oil. Secondly, the process of extracting shale is similar to that used in strip mining. Huge earth-moving machines must first strip away billions of tons of earth to uncover the layer of shale below the surface. This brings with it all the devastation of strip mining. Thirdly, more than 70,000 tons of shale may be processed each day. The by-product of the process is pulverized rock, which in its new form is about 12 percent greater in volume. Where does one store all this material? Mountains of it will accumulate. Occasional rains may wash chemicals from these huge piles into local water supplies, creating a serious pollution problem. Oil shale development may seriously increase the already high salt content of the Colorado River, for example.

Since oil shale covers an area of 8 million acres, a lot of damage to natural surroundings can occur unless extreme care is taken. But the oil is there and the United States is in dire need of it, so a shale industry will certainly appear in the next five years. The same potential and dangers exist for the tar sands of Utah, which contain oil.

Wherever oil exists beneath the U.S. mainland, in the shale deposits of the western states or below the U.S. continental shelf, it will be found and extracted. What remains to be seen is how efficiently we use this precious resource.

GAS

It is rather startling to realize that natural gas, the "modern fuel," was known to the citizens of ancient China as far back as 900 B.C., although we don't know whether or to what extent the Chinese used it. We do know that by the tenth century A.D. about 1100 gas wells had been drilled, the gas being piped to the surface with bamboo tubes. But in the West, and more particularly, in the United States, it was well into the nineteenth century before gas was used on a large scale.

In 1821 the town of Fredonia, New York, was the first to be lit by natural gas. Gas was readily accepted because it burned brightly and without the fumes given off by kerosene or coal oil. But the use of natural gas did not spread rapidly, mainly because there was no practical way to transport large amounts over long distances. That problem was finally solved in the early decades of the twentieth century.

The 1920s saw the development of long-distance gas pipeline systems, which eventually grew to a massive network measuring 223,700 miles, enough pipe to circle the earth almost nine times. Gas consumption took off. By 1946 the gas industry was producing almost 5 trillion cubic feet (5TCF). Public utilities began to extol the virtues of natural gas as the clean modern fuel. The consumer was enticed into buying all sorts of gas-based home appliances—dryers, water heaters, home heaters, even refrigeration units. Consumption rose rapidly. By 1955 the United States was producing about 10 TCF, in 1965 almost 17 TCF, and by 1971 we were consuming over 22 TCF of natural gas annually. This represented almost half the world consumption of natural gas. Then the big squeeze came. Utilities could not keep up with the accelerating demands of this highly prized, relatively nonpolluting fuel. The big pitch to customers to switch to gas suddenly ended because utilities could not handle any new business. The Washington Gas and Light Company of Washington, D.C., for instance, had to reject all new applicants for gas heat service because of the fuel shortage. This was not an isolated case; it has been happening all over the country. Some companies even found it difficult to satisfy current requirements. One answer was to buy natural gas from the Middle East. A number of major companies bought $10 billion worth of gas from Algeria, to be delivered between 1972 and 1997.

Our consumption of natural gas has held pretty steady in recent years —in 1972, 1973 and 1974 it was around the 22-TCF level. This valuable resource presently provides about one-third of our energy requirements. At the current rate of usage, we have only an 11-year supply in reserve. Presently, the United States is using natural gas twice as fast as it is finding it, and the serious supply shortage will inevitably lead to dependence on imports. So far, we have had to import only small amounts, but those amounts are growing. In 1972 we bought less than 5 percent of our natural gas from foreign countries. That figure doubled in 1973, and the demand for natural gas above U.S. production capability could go as high as 25 percent by 1980. By 1990 we might have to depend on foreign nations to supply us with 17 TCF of natural gas each year. Dozens of giant tankers, costing over $50 million each, will carry huge loads of liquid natural gas

(LNG) to our shores. Fortunately, it may be possible to avoid this situation.

One possibility is to increase gas exploration and drill more wells. Theoretically, there is a good deal of natural gas around. Known reserves in the Texas Panhandle amount to about 72 TCF. The U.S. Geological Survey estimates that the total supply of natural gas in our country is about 6600 TCF—enough for 300 years at the current rate of use. However, this estimate includes offshore supplies, which are difficult to extract. Also, the rate of gas consumption is expected to increase steadily: consumption could rise to 40 TCF by 1980. When all aspects are considered, the United States probably has a potential eighty-year supply of natural gas. But we are fighting time; it takes seven to ten years to find, develop and market a new gas find.

One roadblock to increased gas supplies seems to be money. Petroleum and gas companies claim that the price of gas, which is controlled by the federal government (Federal Power Commission), is much too low. In 1973 the average price was 22 cents per 1000 cubic feet. At that price it is not worth investing millions of dollars in drilling. The price of gas will probably have to be allowed to rise significantly to encourage exploration and development of new gas fields. It also costs a great deal to transport gas over long distances. The cost for one proposed gas line, which would run from northern Alaska, through Canada, to join the existing gas network in Montana, is about $5.7 billion. So it will take time and money, two things we are not assured of, to drill for the gas we need. But drilling is not the only way to get gas. Current research on producing gas from coal indicates that coal gasification will produce a high-grade fuel that can be pumped through pipelines along with natural gas by the early 1980s.

Although the major portion of natural gas is consumed as fuel, many industries rely on it heavily as a raw material (feedstock). Industries such as food, paper, chemicals, plastics, petroleum refining and primary metals are consuming increasing amounts of gas for nonenergy purposes. In the chemical industry natural gas is used as a feedstock for ammonia, methanol, acetylene and hydrogen. Even more critical is the fertilizer industry's dependence on natural gas for ammonia production. Ammonia is required to produce nitrogen fertilizers, and fertilizer is absolutely essential for high-level food production in the United States and around the world. Since natural gas is such an important feedstock for fertilizers, it doesn't make sense to burn it when there are so many other methods of generating heat.

As we begin to run out of gas, the price of gas will rise, and this will affect the prices of chemicals, plastics and fertilizers. A serious shortage of gas could really hurt the U.S. economy. It would be in our best interests (and the world's) to conserve natural gas for the most critical applications.

COAL

Spread across thirty states are vast deposits of coal—so vast, in fact, that even if we were to consume this black fuel at the increased rate projected for 1980, we would not exhaust the supply in a thousand years. About 1.5 trillion tons of the stuff awaits mining. Some of it lies just below the earth's surface. Other coal deposits are buried thousands of feet beneath the ground.

Coal has played a major role in industrial history. It was the catalyst and driving force behind the accelerated industrial revolution that took place in England during the eighteenth century. By 1700 England was mining well over a million tons of coal annually. The wood supply had dwindled and coal took its place as a fuel. Coal became the source of energy for home heating and cooking and for industry. Blacksmiths, coppersmiths, pewterers, armorers and other metalworkers made good use of its high heat content. Coal was consumed in large quantities by the brick, glass, tile and earthenware industries. It was even employed in the infant chemical industry in the refining of saltpeter, alum and copperas, for the manufacture of gunpowder, and in the boiling of starch, soap and sugar.

By the turn of the nineteenth century the United States was hauling (on coal-burning trains) the huge amounts of coal required by the steel, chemical and electric power industries. Coal was also consumed in great quantities by thousands of companies for the production of "town gas," the gas that fueled the old-fashioned street lamps and home gas lights. King Coal continued to flourish until the 1920s.

In the twentieth century oil and natural gas began to make inroads into the fuel market. Trains were pulled by diesel-fueled engines, and natural gas, along with electricity, eliminated the need for town gas. The demand for coal began to taper off. It continued to drop until it hit a low point in 1958 of 400 million tons. At this point gas was the fuel of choice for home use: it was clean, maintenance-free and, best of all, cheap. The huge electric utilities were switching to oil because it produced less pollution, and the heralding voices of atomic power predicted vast amounts of clean cheap power. Coal was the fuel of the past, unfit for a modern, technologically advanced society. Then, all of a sudden, the entire picture changed.

In the 1960s we had the first hints of an impending energy shortage. The message that oil and natural gas supplies were not unlimited began to appear in the speeches of government and industry leaders. Furthermore, atomic power had not been developed at the expected pace. Coal as a source of fuel started to receive more attention. By 1965 production was back up to 512 million tons, still short of the production level of 1947, which was about 590 million tons. But in 1970 the coal industry, after a twenty-three-year slump, produced a record-breaking total of 602 million tons. The new demand was initiated mainly by the large electric power plants. By 1974, 400 million tons of coal produced 45 percent of the nation's electricity.

That year the Arab oil boycott shocked us into the realization that our supply of one of the most fundamental elements of our economy—energy —was at the mercy of foreign powers. Because of this trauma coal has an extremely bright future. The projected U.S. consumption in the mid-1980s is about 1 billion tons annually, and by the year 2000 we may be burning 1.5 billion tons of coal each year.

What makes coal so attractive is not its fuel properties but its potential as a raw material for the production of synthetic gas and oil, which are needed by dozens of major industries. Research in coal gasification and liquefaction got a big boost from the U.S. government and industry when, with a budget of $50 million in 1974, the Office of Coal Research, in conjunction with the American Gas Association, made big gains in the construction of three coal gasification pilot plants. It is hoped that the experience gained in the operation of these plants will lead to full gas production facilities by 1980.

The idea of extracting gas from coal is far from new. There was a time when almost every city in the eastern half of the United States had a gashouse where town gas was produced for street and home lighting. In fact, the first American gas company was chartered in Baltimore in 1816, four years after a gas company was established in London. By the mid-1920s there were 150 manufacturers of gas producers around the world. In the United States almost 12,000 gas producers were in operation, consuming about 25 million tons of coal each year. Virtually all of them disappeared with the advent of natural gas. But gas producers did not disappear in other parts of the world. Coal is being gasified today in Turkey, India, South Africa, Scotland, Morocco, Yugoslavia and Korea.

The old time-tested system of coal gasification is known as the Lurgi process. Sasolburg, South Africa, has had a plant operating 13 Lurgi gas producer units for a number of years. This plant provides synthesis gas, which is used in the production of 5000 barrels of synthetic gasoline per day.

Figure 1 illustrates how the Lurgi process works. Coal, which has been crushed to a special size (roughly an inch in diameter), is placed in an airtight pressurized vessel called a producer. The producer is heated to about 900° F. At this temperature coal begins to decompose chemically and the gas it contains expands and escapes. As the coal is being heated, steam, and air or oxygen are forced into the chamber. These elements react with hot coals to form additional gas. A number of different gases are produced in the heated chamber: hydrogen, carbon monoxide, and a small amount of methane. Other gases, such as nitrogen and carbon dioxide, are also formed and must be removed in a later step.

As the coal is used up, the ash is allowed to fall through a rotating grate to the bottom of the producer. What to do with the millions of tons of

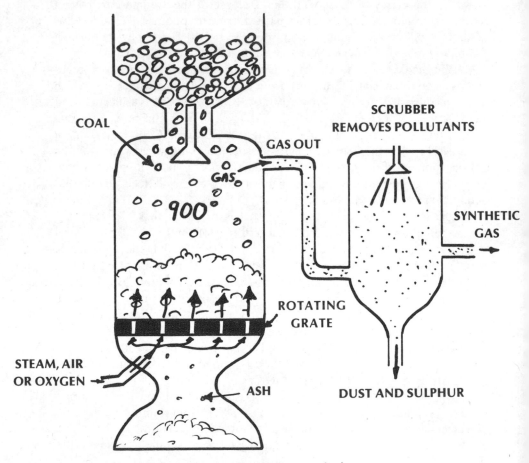

FIGURE 1. The Lurgi process used to produce synthetic gas.

fly ash produced each year is a question often asked by people concerned with the environment. A number of uses have been suggested, among them land fill (topsoil would have to be placed over the ash, which will not support plant growth) and the manufacture of cinder blocks for construction.

The gas generated in the Lurgi process is of low heat value—in other words, when it is burned, it does not produce a great amount of heat. This producer gas generates about 500 BTU as compared with 1000 BTUs from the natural gas we presently use in our homes. Furthermore, it contains certain impurities such as sulphur and small dust particles, so it must go through additional steps before it is considered usable.

First, the gas is cleaned. It is sent through a "scrubber" chamber where the sulphur and dust are washed out. (The sulphur can be refined out and sold for its commercial value.) The next step in the process increases the heat value of the synthetic gas to a level that will allow it to be mixed with natural gas in the nation's vast network of gas pipelines. In this phase of the process (methanation) the gas, which contains carbon monoxide (CO) and hydrogen (H_2), is passed over a very hot metal (nickel) called a catalyst. The catalyst causes most of the CO and H_2 to combine to form methane (CH_4). Natural gas is almost pure methane, and methane burns with a very hot flame.

The Lurgi process is one of the few commercial methods presently available (another method, the Koppers-Totzek process, is being used in Europe), and it has some limitations. First of all, the system is not capable of producing large volumes of pipeline-quality gas. Then there is the problem of coal size. The Lurgi producer requires coal of a specific size (from $\frac{1}{2}$ to 2 inches in diameter). It cannot process very fine grain coal, such as that produced from strip-mining techniques. (The Koppers-Totzek process solves some, but not all, of these problems.) The U.S. Office of Coal Research and the large power and fuel companies are pressing hard to develop an efficient high-volume gasification process. The stakes are high and the rewards could be enormous. What they are aiming for is not just a coal gasification process but a system that will produce coke (which is used in steel), oil and natural gas—all from coal, which we have in abundance. Huge COG (for coke, oil and gas) plants could allow the United States to reach energy self-sufficiency by the mid-1980s.

Certainly the first stages of coal gasification could be completed by 1978 or 1979. Three coal gasification pilot plants have already been built. A larger commercial demonstration plant, costing about $200 million, is the next step. The current outlook is that by 1990, 45 large plants could be processing up to 300 million tons of coal to produce about 3.5 TCF of gas. That would be enough fuel to supply approximately 10 percent of U.S.

energy needs. In one form or another, coal will be in the U.S. (and the world) picture for many decades.

SHORTAGES ARE ONLY PART OF THE PROBLEM

Long gasoline lines and higher fuel prices may make our lives a little more difficult, but they are by no means the most serious consequences of the energy crisis. The potential problems from the energy shortage vary greatly in nature and in scope.

Our basic approach to energy must be reexamined. For example, we continue to think in terms of how long our oil, gas and coal supplies will last as sources of fuel. This in itself is a mistake. Using these precious fossil materials to generate heat is preposterous when the sun is continuously showering the earth with trillions of calories of heat, the oceans are a vast store of solar heat and the huge core of the earth is an inferno capable of providing energy to mankind for millions of years. We should reserve our valuable store of hydrocarbon materials for nonenergy uses—as raw materials for plastics, lubricants, medicines, synthetic fibers and thousands of other products necessary to our society. If we did this, they would last for thousands of years instead of the 50 or 60 years now projected.

Then there are the political consequences of the energy situation. During the years the United States was producing more fuel than it consumed, we were completely independent of foreign suppliers, and even able to supply other nations with significant amounts of fuel in times of crisis. This situation began to change in the early 1960s, and today, with oil and gas reserves falling, we depend heavily on other nations for the fuel that is essential to our industries and our defense. Our imports of petroleum have increased dramatically. In 1965 the United States consumed about 12 million barrels of oil each day; we produced 9 million barrels domestically and imported about 2.5 million barrels (mainly because of price). But we had a spare production capacity of 3 million barrels per day. By 1972 we were going through 16 million barrels per day (a 33 percent increase over 1965) and producing 11 million barrels. Thus we had to import over 4.5 million barrels per day. And we had no spare capacity. We had become an energy-dependent nation.

By 1985 we may be consuming about 25 million barrels of oil per day. Since our production capacity may be less than 12 million barrels per day, we will have to import almost 14 million barrels of oil per day. The ramifications of this dependency are numerous. For one thing, our balance of payments (the amount of money flowing out of our country as opposed to the amount of money flowing in) could be disrupted. If we are forced to import 14 million barrels of oil per day, at $10 per barrel, we will be footing a bill of $140 million each day, or over $50 billion annually. The situation may be even worse for Japan and the other oil importing nations of Western Europe that have fewer or no oil reserves.

It is no longer possible for the United States to bail out its allies with large amounts of fuel. In fact, we are now in direct competition with our own allies for fossil fuels. For example, in 1972 Japan and Western Europe consumed about 2.5 million more barrels of oil per day than the United States did, and their needs required 69 percent of all the oil produced in the Middle East and Africa. This competition for fuel can only become more severe, and one of its consequences will be to exacerbate relations between the United States and its allies. After the Arab-Israeli war of 1973 Japan, for example, was forced to change its foreign policy to favor the Arab nations at the expense of its friendly relations with Israel because it was dependent on Arab oil for its survival. As fossil fuels become more scarce, other of our allies will feel compelled to make similar changes and the traditional alliances the United States has depended on may be upset.

POLLUTION

Sulfuric acid is a dangerous chemical. It must be handled with extreme care. Even a drop falling on one's skin could cause a very painful burn. Each year upwards of 400 billion pounds (counting water content) of sulfuric acid are dumped upon the surface of the earth, mainly as a result of burning fossil fuels. This is only one of the many forms of pollution resulting from our present approach to energy generation.

Almost all of this acid (90 percent) falls on the northern half of the earth. "Acid rain," as it is called, sinks into the ground, stripping away or leaching precious calcium from the soil. Since tree growth is correlated with the calcium content of the soil, trees suffer. According to Dr. Ian C. Nisbet of the Massachusetts Audubon Society, tree ring studies in Norway have revealed a definite decline in growth. And here in the United States acid rain is threatening the health and vigor of the New England forests.

Acid rain also affects fish. In Norway and Sweden, which receive their pollution from the industrialized countries to the south and west, the rivers are becoming acidic, so much so that fish such as salmon and trout are ceasing to breed. If the acid level continues to rise, many of the fish will die. Nor is North America being spared. Fish kills have been reported in acidified lakes in northern Ontario, Canada, and in some parts of New Hampshire the rain has a very high level of acid.

But the problem goes even further. Dr. Gene G. Likens of Cornell University and Dr. F. Herbert Bormann of Yale state that acid in the atmosphere is causing serious damage to bridges, outdoor stations and other exposed structures. The cost for structural repairs and replacement is huge.

Pollution comes in many forms and affects the ecological systems of the world in countless ways. Air pollution is perhaps the most familiar. We have all heard or read about the devastating London black fog, which directly or indirectly resulted in the deaths of many citizens, especially older people and those with respiratory conditions. Los Angeles has taken the brunt of many jokes because of its smog (a polluted form of fog). But air pollution is no joke. It destroys structures, vegetation and people. It is of prime importance since we are constantly in intimate contact with this portion of our environment.

Among the more dangerous constituents of air pollution is a colorless, tasteless, odorless gas that is a merciless killer—carbon monoxide (CO). Because of some quirk in nature, carbon monoxide has a greater affinity (200 times greater than oxygen) for the hemoglobin molecules in the red blood cells of our body. When we inhale CO, it crowds out the oxygen molecules that normally attach themselves to the red blood cells in our lungs. Therefore CO starves our body of its important oxygen supply. Significant amounts of inhaled CO can cause serious illness and even death. In a study carried out by the American Medical Association it was found that a great number of Americans are inhaling carbon monoxide in quantities far above those normally considered safe. This poses a serious threat to health, especially for persons with heart conditions whose blood oxygenation system is already impaired.

Considering the huge quantities of pollutants that are belched into the atmosphere, it's little wonder that the rate of respiratory disease continues to rise. U.S. smokestacks discharge close to 300 billion pounds of pollutants, including 29 billion pounds of sulphur dioxide, 360 million pounds of lead and billions of pounds of particulates (solid particles) such as fly ash. To take just one example out of hundreds around the country, the Four Corners Power Plant in Fruitland, New Mexico, spews 250 million pounds of fly ash into the air each year. Particulates get into our lungs, causing congestion. Some particulates may be directly toxic to man (beryllium, arsenic). Others (tars, asbestos) are definitely carcinogenic (can cause cancer). Because particulates can absorb gaseous pollutants (such as sulphur dioxide), they are able to penetrate deep into the lungs and cause disease.

The devastating pollution of our water aggravates the situation. Water pollution can result from the transfer of air pollution elements such as sulphur dioxide or particulates into lakes, rivers, streams and oceans. It can also come from oil drilling rigs, pipelines, storage tank leakage and shipping accidents. Large oil spills, such as the one off Santa Barbara, California, or off the Massachusetts coast in Buzzard's Bay, can seriously damage the aquatic environment, destroying wetland spawning areas and recreational beaches. Large oil tankers continue to clean their bilges on the open seas, adding to the amount of oil that is floating on the surface of the water. Because of all this accidental and deliberate pollution, even the most remote portions of the oceans show evidence of oil contamination.

Adding to this already serious condition is the water pollution resulting from coal mining. Our once beautiful streams and lakes are being degraded on a grand scale. The Department of the Interior estimates that by 1973, 13,000 miles of streams and 145,000 acres of lakes and reservoirs had been adversely affected by acid mine drainage and siltation from coal extraction.

Heat is the final waste product of all uses of energy; whether one is driving a car, making toast, lighting a room or shaving, heat is generated. There are billions of sources of man-made heat, and heat is the ultimate form of pollution. It is heat that limits the amount of energy that man may safely generate on this planet.

Aside from making us very uncomfortable, heat pollution can create far more serious problems. It has an effect on the weather: frequency and amount of rain, local temperature, air quality, all are influenced by the amount of heat discharged to the environment. Since the difference in temperatures around the earth is a major factor governing weather patterns, any significant change in heat input and heat concentrations will affect weather conditions. Eventually, the introduction of vast quantities of heat into the atmosphere and water systems may create totally new weather patterns, which could be highly destructive. And this is only one aspect of the heat pollution problem.

A great amount of waste heat is introduced into our streams, rivers and lakes, much of it coming from electrical power plants. The water is used for cooling purposes; that is, heat is transferred from the steam system (which drives the steam turbines that operate the electrical generators) to the cooling water. This water is pumped in from local rivers or lakes. As the water leaves the power station, it is between 11° and 16° F. warmer. This heated water is then transferred to the river or lake. It takes a lot of water to cool the power plants of the United States. In 1959 over 26 trillion gallons were used for this purpose. By 1970 the figure jumped to 57 trillion gallons. It is predicted that by 1980, 106 trillion gallons of water will have to flow through power plants to keep them cool. To appreciate this alarming situation, one need only look at the demands for cooling water predicted for the turn of this century. The amount of water required to cool all the electrical power plants that will exist in the year 2000 will equal all the fresh water runoff (resulting from rain) of the entire continental United States for a full year. Furthermore, fairly large areas of water are directly affected by heat pollution. Even by 1980 the electrical power plants of the United States will have to discharge 14 trillion BTUs of waste heat. That's a lot of heat even for nature to handle. Incidentally, this wasted form of energy is equivalent to more than 2 billion barrels of oil, or about 20 percent of all the oil that the entire nation will consume by the end of this decade. The thermal discharge from a 1000-MW megawatt) plant will affect 8 to 10 miles of an average-sized stream or about 5.5 square miles of a lake or bay. What does this heat do?

One damaging result of an increase in temperature is that a body of water loses its ability to hold oxygen. This reduces the water's natural ability to cleanse and purge itself of accumulated wastes. Elevated water temperature also has an effect on aquatic life. Scientists claim that as water temperature rises, many fish require more oxygen. Since heated water holds less oxygen, these fish suffer loss of appetite, digestive problems, difficulty in breathing and reduced rates of reproduction.

Considering the effects of waste heat on the earth's atmosphere and on its bodies of water, it is clear that we face some serious challenges as energy generation and consumption increase rapidly. In fact, heat pollution and the ultimate heat radiation capability of the earth may very well be the most powerful reasons for our society to turn to other, more efficient systems of generating energy. Instead of considering the cost of energy only in terms of fuel price, labor and capital expenditures, we should also estimate how much waste heat a system adds to the environment and assume this is part of its cost.

What can we do about the present heat problems? Nothing can be done to prevent the conversion of used energy into heat. But distribution and methods of discharging heat can be improved. A local lake or stream might be seriously damaged by excessive heat absorption, whereas the same amount of heat discharged into the air might have a less dangerous effect. Electrical power utilities are incorporating a number of systems to disperse waste heat in the least damaging manner. Artificial lakes have been constructed to supply cooling water. The heat is then evaporated to the atmosphere. But where land is very costly, cooling lakes can be expensive. A 1000-MW electrical generating plant, which will use 500,000 gallons of cooling water per minute, would require an artificial cooling lake covering 2000 acres. In the Northeast and on the West Coast this much land is almost prohibitive in price. Cooling towers (called hyperbolic towers) that rise 400 feet into the air are another method of discharging heat from cooling water. But they are also expensive.

In other areas of pollution new equipment and new methods are being developed. Various systems that remove particulate matter and chemical effluents from chimney exhausts are being installed. More efficient methods of burning fossil fuels that reduce pollution, such as fluidized-bed combustion, are being introduced.

But pollution-control equipment is costly. Electrical power utilities spent $565 million in 1971 for this type of equipment. One year later the cost jumped 82 percent to $1 billion—the largest expenditure for pollution-control equipment of any industry. U.S. utilities may have to spend over $48 billion by the mid-1980s for hyperbolic towers used to cool water and for other specialized equipment. Obviously, this means an increase in the average citizen's utility bill.

The entire pollution problem is mind-boggling. Even if we believe the most optimistic projections, it is clear we are faced with distressing decisions. In some instances we don't have a ready solution to the problem. Where we have a solution, it is either very costly or demands sacrifice or a basic change in our life-style. Above all, we must change our attitude concerning the real cost of energy. Fuel, labor and capital equipment are three major ingredients of the cost of energy, but the overall price our society pays for energy must also include the cost of damage caused by pollution—repairs, cleaning and replacement of buildings, bridges, statutary and other structures. More important, it has to include the huge medical expenses incurred by citizens who suffer from the effects of pollution—heart patients, people with respiratory diseases, eye damage or bad nasal conditions. A significant portion of these medical bills can be directly attributed to pollution.

Therefore in choosing future methods of generating energy, especially electricity, we must not eliminate systems such as solar, wind or geothermal power because they do not compete economically with our present systems in terms of cost per kilowatt-hour. Considering the cost of pollution, considering the health of our citizens, considering the long-range fuel supply and the potential danger to our ecological systems, it may be far more profitable to us and to future generations to choose more natural methods of generating energy.

SOLVING HALF THE PROBLEM: CONSERVATION

To survive at our present level of energy consumption we will have to take strong measures on two levels. Half the problem is finding better, more efficient ways to generate huge quantities of energy without disrupting nature's ecological arrangement. This will be the subject of subsequent chapters. The other half of the problem has to do with wasting energy. It may appear the lesser part of the problem but, in fact, energy waste is probably the biggest reason why we are presently suffering from a lack of certain forms of energy. If we are to overcome the energy crisis, we will have to practice conservation.

The way we use energy today is a direct result of a long history of plentiful energy. Our use of big gas-guzzling cars and trucks instead of more efficient rail and waterway transportation, our construction of huge glass office buildings requiring tremendous air-conditioning units—these are examples of our careless assumption of continued abundant cheap energy. A well-planned conservation program could reduce this waste dramatically. According to Eric S. Cheney, associate professor of geological science at the University of Washington, we should be able to reduce our energy consumption by 25 percent without any significant change in our life-style.

There are literally thousands of things that we as individuals and as a society can do to save energy. When the University of Indiana launched a well-coordinated campaign to turn out lights, reduce water dripping from faucets and affect other energy saving procedures, its utility bill dropped $150,000 in six months. Even bigger savings can be attained by industry. Instead of producing steel from raw materials, it can be made by recycling scrap steel, which would save about 70 percent of the energy used in the process. Process improvements, maintenance methods and increased use of insulation can also reduce fuel consumption at steel mills. In fact, an average-sized steel mill can save about $1.3 million in fuel by instituting an energy conservation program. The aluminum industry has the same opportunity. Scrap aluminum can be recycled with 66 percent less energy than is needed to process bauxite. (Of course, a company must consider the cost of collecting scrap aluminum, which can sometimes be greater than the savings in terms of energy consumption.) Not only does conservation save individuals, companies and society money, but it also slows the rate at which our fuel supplies are depleted and it helps to reduce pollution.

Conservation must be practiced in all sectors of our economy. Air transportation is one area where we could reap large savings: studies have shown that reducing commercial airline cruising speeds by 3 percent would add only 12 minutes to a cross-country trip, but would save the United States 200 million gallons of jet fuel per year. Furthermore, we could reduce many of the half-empty flights that lose money for the airlines and waste a lot of fuel; if by reducing flight frequencies we could raise the national aircraft load factor from 52 to 60 percent, we could save approximately 1.3 billion gallons of jet fuel each year.

Some scientists have suggested a return to giant airships. Such ships, similar to the dirigibles of the 1930s, would be ideal for carrying large amounts of freight across the country; each airship could carry upwards of a million pounds of cargo. Although not fast (about 70 to 90 mph), the ships would be safe, since they would use inert helium gas. Most important, airships could save billions of gallons of jet fuel.

Similar approaches for saving energy could be applied to other industries from paper production to plastics. "The fact is," says Russell E. Train of the U.S. Environmental Protection Agency, "industry in general could probably reduce its energy consumption almost 35 percent by 1990 with a concerted effort." But industry is not the only energy-wasting culprit.

The commercial sector could make a good deal of improvements in its use of energy. Take our huge office buildings. Towers of steel and glass, they are fully heated in the winter and fully cooled in the summer. That represents a great amount of energy, one-third of our total consumption, in fact. While they may look ultramodern, our glass towers are highly inefficient when it comes to energy. Glass transmits heat at five to ten times the rate of insulated walls. In a modern office building up to 30 percent of the energy used on air conditioning is wasted because of large glass windows. According to Richard Roberts, director of the National Bureau of Standards, as much as 40 percent of the energy expended by commercial buildings is wasted. Nationwide, this loss is equal to about 65 billion gallons of oil each year. But architects could design more efficient buildings. Roberts gives a good example. The architects for the Toledo Edison Building selected a chromium-coated dual-wall insulating glass that cost $122,000 more than ordinary plate glass—a lot of extra money just for glass. But since the energy saved by the glass will amount to about $40,000 a year, after four years the owners will be ahead of the game. Considering the increasing prices of fuel and electricity, in ten years they will have saved well over $400,000. Conservation can pay very handsome dividends. Energy can also be saved in the general maintenance of buildings by not fully heating and cooling large lobbies, by making sure that outside doors shut automatically, by reducing the number of lights and by keeping all energy-consuming equipment in peak operating condition.

Conservation of energy must become a major factor in the development of new products. The computer industry is a great example of how energy consumption can be drastically reduced. The Hewlett Packard HP-70 hand calculator can perform as many functions as could the original Eniac computer built in 1948. Yet it consumes millions of times less power than did the Eniac, which filled an entire room with vacuum tubes and other electronic parts. This remarkable difference was made possible by an invention known as the integrated circuit, which packs thousands of electronic parts on a silicon chip only a quarter of an inch square. The complex integrated circuit that performs all the computer functions consumes only a fraction of a watt of power.

Industry and commerce are not the only energy wasters on a grand scale. The 70 million homes we live in are excellent places to apply the rules of energy conservation. Our homes account for 20 percent of all the energy consumed in the United States, and we could reduce our consumption of energy there by about 25 percent without any radical changes. For starters, it wouldn't hurt to turn down the thermostat a few degrees. Doctors claim that it is even a bit healthier to keep temperatures slightly lower than those we are used to. We might try mowing the lawn by hand instead of following a gas-guzzling or electricity-consuming machine around the property, using nonelectric blankets and brooms, cutting down on extensive use of the large-screen color TV. These practices may seem trivial, but besides saving some energy, they would help put us in a better frame of mind regarding energy. And a big part of the crisis is our mental attitude; we're just not energy-conscious, though we are fast becoming so as our energy bills increase almost monthly.

Just consider that of all the energy generated in this country, only 40 percent is used for productive purposes; the other 60 percent is wasted. One place where it is wasted is in the home. More than half the heat in our homes is lost during the winter through the effects of infiltration—that is, the entry of cold air through drafts, open doors, poorly caulked windows, etc. Since heating can account for half of all the energy consumed in a home, conservation of heat can result in a large savings in fuel, and in reduced fuel bills. Lowering the thermostat 4° F. in a single-family house can save the equivalent of 158 gallons of oil per year. Adding sufficient insulation in roofing, attic and around hot-water pipes is another way to save on fuel consumption. Six inches of insulation in the attic can save the equivalent of 244 gallons of oil in a house with an oil-fueled heating system.

In the summer a fan can be made to do the important job of cooling a house. Hot air rises, so an attic may be 40° F. hotter than the rest of the house. An inexpensive fan ventilating the attic can do wonders in keeping the house at a reasonable temperature. It takes about 300 kwh (a kilowatt-hour = 1000 watts for one hour) annually to run an attic fan. A central cooling system in a 1500-square-foot house may take 4000 kwh.

Home appliances are another source of energy waste because most of them are inefficient. Since we have so many appliances around the house, they can help us add to our energy savings. One way is to use them less. Dishes dry perfectly well at room temperature, so why put the dishwasher through its drying cycle? The quality of appliances is also important. Often higher-priced devices save money and energy in the long run. A Harvard study compared two refrigerators of equal size. One cost $292 to buy but would cost $746 to operate over its 20-year lifetime. The second one cost $359 ($66 more) but only $392 to operate over 20 years. So the higher-priced refrigerator is $287 cheaper to own. The same is true for other appliances.

The family car is the biggest energy-consuming piece of equipment owned by the average American. In 1970, 46 percent of the world's passenger vehicles were on U.S. roads and highways and passenger cars accounted for 21 percent of all the energy consumed in the nation. Unfortunately, the family car is not a very efficient machine. Consider that the energy expended by a person riding a bicycle one mile is about 180 BTUs, but an automobile consumes about 4250 BTUs traveling one mile. The car has been singled out by many authorities as the worst offender in the energy crisis. According to William Ruckelshaus, former administrator of the Environmental Protection Agency, "Sheer bulk is the worst offender: an increase of only 500 pounds in car weight—say, from 2500 to 3000 pounds—can slash mileage from 16.2 miles per gallon down to 14 miles per gallon, a drop of 14 percent. A 5000-pound vehicle consumes 100 percent more gas than does its 2500-pound counterpart." Mr. Ruckelshaus goes on to say that a drop from the present average weight of the automobile to a 2500-pound maximum would reduce total gasoline consumption in 1985 to the projected level of 1975. This would reduce crude oil imports by 2.1 million barrels per day in 1985.

The automobile, then, is excellent for practicing conservation in a number of ways. One might change from a regular to a compact car, which uses about 33 percent less gasoline. A car should also be kept well tuned; a badly tuned engine can reduce gas mileage considerably. Reducing driving speed can also save energy. A car driven at 75 miles per hour will consume almost 100 percent more fuel than the same car driven at 50 miles per hour. Car pooling can also pay off. If a person who commutes 30 miles a day, round trip, joins a three-car pool and drives to work only once every three days (in a full-sized car), he or she would save about $216 in fuel and the equivalent of 590 gallons of crude oil each year.

Another way of conserving car energy is simply not to use the car as often. Instead of making a dozen trips to different stores, one should plan ahead and buy all or most of the items on one trip. Where there is a mass transit, it should be used. Buses and trains (including subways) are far more efficient in terms of passenger miles (1090 BTUs and 1700 BTUs versus 4250 BTUs for a car). Since buses and trains must operate whether or not they are filled, it is important in terms of conservation that people use mass transit as much as possible. If one-half of the intercity air traffic and one-quarter of the intercity auto traffic could be switched to passenger trains, and if railroads operated at 70 percent instead of 25 percent capacity, the United States could save about 11 billion gallons of gasoline each year.

Conservation can be practiced not only in the use of things, but also in their design. Most appliances such as refrigerators, freezers and air conditioners could be designed to last longer (thereby conserving the energy that went into producing them) and to operate more efficiently. The incandescent light bulb is a good example: it converts only 10 percent of the electricity it uses into light; the other 90 percent is converted into heat. This waste of energy is not too bad in the winter when heat is welcome, but in the summer these lights add unwanted heat and cause the air-conditioning system to work harder and consume more energy. Some household conveniences cost a great deal more than is realized. For example, a frost-free refrigerator uses over 60 percent more electricity than its conventional counterpart, and a gas pilot light uses 50 percent of the total gas consumed by a stove. Actually, pilot lights are particularly wasteful because inexpensive automatic devices can ignite a burner conveniently and safely.

There are, however, sources of energy waste and inefficiency that affect our present crisis on a much larger scale. Electrical power utilities are very much in need of significant design improvements. Much environmental damage comes from systems that provide energy to consumers. If these systems functioned more efficiently, these adverse environmental effects would be reduced. Improving the efficiency of a power plant is even more important than installing pollution controls at a utility. Efficient systems burn less fuel, and therefore reduce not only the amount of pollution at the power plant but also the pollution factors associated with extraction, processing and shipping the fuel.

Electrical generating systems range in efficiency from a low of 10 percent to a high of only 25 percent. So even with the most efficient system presently operating, 75 percent of the energy is wasted, usually in the form of heat. Waste heat represents a terrible loss of energy for the United States—if only 10 percent of the heat dumped from power plants were used in the next 30 years, it would equal today's U.S. electrical output. Right now we are dumping enough to heat all the homes in America. In addition, we are heating a significant percentage of our freshwater supply: the equivalent of 16 percent by 1980 and possibly 100 percent by early in the next century.

The surprisingly low efficiency of electrical generating systems results from the loss of efficiency all along the line in the production of electricity. Right at the start of the process a certain amount of energy is expended in deep-mining and surface-mining coal extraction processes, and in drilling for oil or gas (70 percent, in the case of onshore oil drilling).

Once the fuel is extracted, it must be processed. Processing can be anything from simple procedures such as crushing and washing coal to the highly complex methods required to prepare uranium for atomic reactors. The production of electricity also consumes a certain amount of energy and further reduces the efficiency of the total process. Then a small amount of energy is consumed in transporting the fuel to the location where it will be used.

The cumulative effect of all these losses of energy is a very low total system efficiency. For example, the system that delivers electricity to our homes (based on onshore oil), is only 10 percent efficient. We have already noted that the light bulb is only 10 percent efficient—and since this is 10 percent of a system that is itself only 10 percent efficient (the onshore oil-based system), the light bulb represents a total efficiency of only 1 percent. So for every 100 watts of potential energy in the original fuel, only 1 watt is converted into light; the other 99 watts are discharged as waste heat. And this is under normal conditions. On a hot summer evening an air conditioner, which uses a considerable amount of energy, may be required to dissipate the waste heat from the room. Therefore it is possible for a system to waste more energy than the total potential energy of the original fuel.

How do we get ourselves out of this appalling situation? One method of improving our present electrical generation system is called the combined cycle, in which various systems or cycles are combined to generate electricity. Figure 2 shows the conventional steam cycle used today by many utilities. Water is boiled in a tank by burning oil, gas or coal. The high-pressure steam then rotates or drives a steam turbine, which in turn operates on an electrical generator. After the steam passes the turbine, its work is done, and it is returned to the water tank. But first it must be cooled so that it will change back to water. This is done by passing the steam through a device called a condenser. Cold water from a lake, river or stream is pumped into pipes that flow around the condenser and the heat is extracted from the steam. This heat is then carried out to the body of water, where it eventually dissipates to the atmosphere. The lower part of Figure 2 is a simple flow chart that shows the basic action in a stream-cycle system.

Another form of generating electrical power used in the past ten years involves a type of jet engine called a gas turbine. As shown in Figure 3, air is sucked into the engine by an air compressor. It then enters a burner chamber, where fuel is added to the air and ignited. The hot (1000° c.) gas leaving the burner chamber passes through a gas turbine connected to an electrical generator. Notice that the flow chart for the gas turbine is practically identical to that for the steam cycle. The main difference (other than the fact that one uses a steam boiler and the other uses a gas turbine) is that the waste heat leaving the gas turbine is much hotter than the waste heat from the steam-cycle system. Usually, all the heat leaving the gas turbine is exhausted into the air, but this heat can be put to very good use.

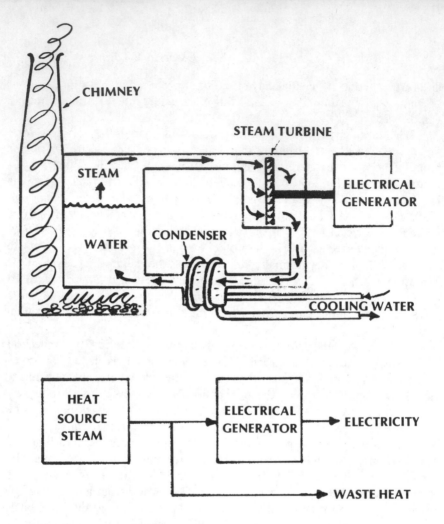

FIGURE 2. Conventional steam cycle for electrical power generation.

Under the combined-cycle system, the two types of cycles are integrated. The hot exhaust of the gas turbine (445° C.) is routed to the boiler of the steam-cycle system. This hot exhaust can produce steam with a temperature of 390° C., and in this way, most of the heat is put to good use. (See Figure 4.)

By the early 1980s a combined-cycle system should reach 50 percent efficiency as opposed to the 35 percent efficiency of today's systems. The steam and gas-turbine systems waste a great deal of heat. The steam system, in fact, requires additional energy to operate pumps, and this extra energy wastes a lot of heat, which simply compounds the inefficiency of the system. Combined-cycle systems can eliminate up to 25 percent of the heat waste.

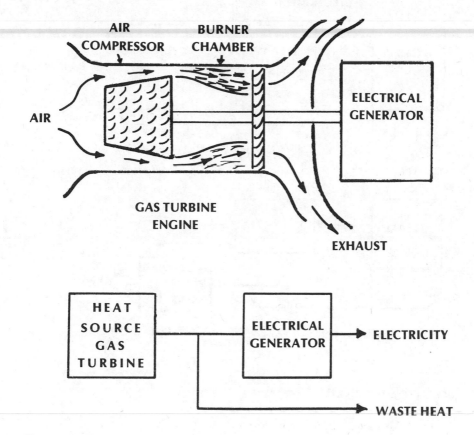

FIGURE 3. Electrical power generation using gas turbine engine.

The added gas turbine (see Figure 4A) is referred to as a topping cycle, because of its higher operating temperature. Similarly, a bottoming cycle can be added to the system (Figure 4B), and in this case a liquid such as isobutane, which boils at a much lower temperature than water, is used. Therefore the heat collected from the steam leaving the steam turbine can be used to heat the isobutane to create steamlike gas. Special turbines are being developed to operate on isobutane low-pressure gas. The bottoming cycle further reduces waste heat and increases system efficiency.

More advanced systems may be developed in which as many as four cycles will be combined for greater efficiency. In the advanced system illustrated in Figure 5, the high end of the system, or topping cycle, uses potassium. Liquid potassium would be heated in a boiler to form a high-temperature vapor that would drive a turbine. The exhaust from the combustion (used to heat the potassium) would be passed through a gas turbine, which would operate a second electrical generator. At the low end the

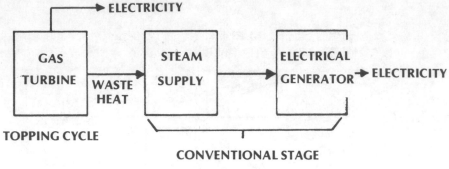

(A) TOPPING CYCLE

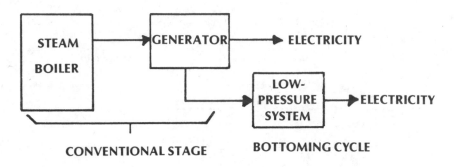

(B) BOTTOMING CYCLE

FIGURE 4. Using higher and lower levels of heat increases the overall efficiency of electrical power generating systems.

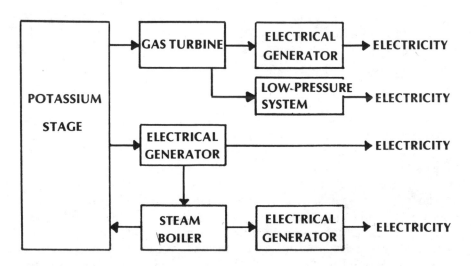

FIGURE 5. Advanced combined-cycle electrical power generating system.

exhaust from the gas turbine would be used as heat for an isobutane low-pressure system. It would run a third electrical generator. Finally, the heat of the liquid potassium system would be extracted by means of a condenser and used to produce steam to run a steam turbine. The turbine would drive a fourth electrical generator. It is not unreasonable to expect such an advanced system to eventually operate at an efficiency of 55 percent.

Combined cycle systems are just a method of using fuel efficiently. Finding enough fuel, and extracting it without disrupting nature's delicate balance and world political stability, remains problematical.

Unfortunately, the first shots of the energy war have already been fired. OPEC (Organization of Petroleum Exporting Countries) is in full control of most of the oil and gas needed by the West. The startling price increases of the last few years are threatening the economic stability of the oil-consuming nations that depend mainly on imports. The balance-of-trade deficits of such nations as Italy, France, Britain and Japan could eventually lead to a worldwide depression. These political conflicts could have been avoided if we had been more flexible in our approach to energy and not stuck rigidly to the use of fossil fuels.

But the political problems are not the most serious. In the face of diminishing supplies and increasing pollution, we still continue to think of oil, gas and coal mainly as sources of fuel, and this hidebound thinking may ultimately cause us the greatest difficulty. It is estimated that there may be no natural gas available for electrical power plants by 1985. Coal will not make a significant additional contribution until the end of this decade. It will reach an annual production of only 800 million tons of 1985 instead of the 1 billion tons required. Nuclear electrical plant output will fall far short of the 1985 requirement. Only 10 quadrillion BTUs will probably be available, while 20 quadrillion BTUs will be needed. The total *deficiency* of domestic sources of energy for the next 12 years will be nearly equivalent to all the energy consumed in the United States in the past 10 years. In short, 125 quadrillion BTUs of energy will be required in 1985, and only 82 quads will be produced domestically. The remaining 43 quads will have to be imported.

It is clear from these figures that we are rapidly running out of fuels. For electricity, we can always turn to atomic power, solar, geothermal and other forms of energy. But what about the other uses of hydrocarbons (fossil material)? Our industries are now producing 10,000 petrochemical products, and Dr. John J. McKetta of the University of Texas (former chairman of the National Energy Policy Committee) predicts that there will be a severe scarcity of raw materials for these industries until the next century.

The significance of this shortage is startling. Just in the plastics indus-try, a study by Arthur D. Little reveals that 2.2 million jobs in industries selling up to $90 billion worth of goods annually depend on PVC (poly-vinyl chloride) which comes from chlorine and petrochemicals.

Perhaps the energy crisis has been a traumatic experience for us be-cause we have such an unrealistic view of energy. Energy, in the form of light, heat, electricity, fuel, has historically been cheap and plentiful in the United States, and we assumed it always would be. Now, suddenly, we are faced with the frightening proposition that we have outrun our ability to generate vast quantities of energy and face a challenge to our survival as a viable industrialized nation. We will be forced to go through a long period fraught with difficulties, economic, social and political. Above all, we will have to develop a national conscience regarding energy and under-take the task of totally reconstructing our energy systems on the basis of sound principles regarding sources, use and economics. Fortunately, there is still time to meet this challenge, if we act fast and decisively. Where do we start? Which direction do we go?

The key to the solution is reliance on energy sources that are self-renewing, universally available and in harmony with nature. The two forms of energy that meet these requirements are electricity and hydrogen. To-gether, they can perform any task required by a modern society.

2

A Clean Break
with the Past

Our wishing well is going modern on us. It is becoming the total electric wishing well with lights, automatic controls . . . it even has a computer to keep up with our myriad requests. Maybe the wishing well is trying to tell us that invisible, silent, nonpolluting electricity can do more jobs for us—and do them better—than any other form of energy on earth.

If we forget its high cost for a moment, it is obvious that electricity is the best possible form of energy for our homes. It is absolutely clean, quiet and practically maintenance-free. What more could one ask of an energy source? The advantages of electricity have been recognized for decades, which is why it is the fastest growing of all our energy sources, doubling in consumption about every ten years. Actually, over the past 75 years our consumption of electricity has grown about 75 times faster than our total energy consumption. Much of that growth has been a shift from gas and oil to electricity in areas such as heating and appliances. Today electricity accounts for 27 percent of all the energy consumed in the Untied States.

Although we tend to associate electricity with modern society, the electrical power industry actually goes back to 1878 when St. George Lane-Fox, an English inventor, proposed a plan to transmit electrical energy to customers for use in lighting. During the same year a young American inventor named Thomas Edison proposed a similar system and the race was on. Fox won by eight months. The first public power station went into operation in London on January 12, 1882; Edison's station started operating on September 4 of that year. They both produced direct current (DC), which is very inefficient, but compared to gas and oil, a light glowing brightly with no smell, fumes or flickering was a great improvement.

Unfortunately, because of its low efficiency the system could not transmit power over long distances, and a single city might require a dozen or more separate electrical power distribution systems. But this problem was solved in 1886 when George Westinghouse and William Stanley demonstrated a new transmission system that operated on alternating current (AC). With this new method, they were able to transmit power over 4000 feet and take the first big step toward large electrical power transmission networks. By 1896 large amounts of electrical power were flowing from the Niagara Falls Hydroelectric plant to the city of Buffalo, New York, a distance of 22 miles. In 1934 power generated at Boulder Dam on the Colorado River was transmitted to Los Angeles, 270 miles away. By 1965 the United States was using almost 1 trillion kilowatt-hours; in 1973 we reached 2 trillion kwh. And so it goes. Projections for the year 2000 are as high as 10.25 trillion kwh, more than 20 times the amount used in 1947. Why the rush to electricity?

First of all, it is quiet. It surges through the house, office building or factory without adding a single decibel to the noise level. Electricity is also the cleanest of all energy forms. Except for heat, which is the end product of all expended energy, electricity is pollution-free. It produces no fumes, gases, particulates, radioactive waste or chemicals to degrade our air or water. A house or a commercial building operating on electricity alone adds virtually no pollution to the environment. Furthermore, except for the need to change a fuse occasionally, an electrical heating system operates for many years without any maintenance requirements, which cannot be said of oil or coal systems. Even gas heating systems require more maintenance than electrical systems.

An electrical system takes less space than any other system; it can be installed in floors or in flat baseboards. Since it does not require fuel, there is no need for a large oil tank for storage or a boiler to produce steam or hot water for heating. Therefore it permits the basement of a home to be free of cumbersome devices, with the exception of a hot-water tank. Another advantage offered by electricity is the elimination of fuel delivery— no cold winter mornings when oil is not delivered, no wondering if the local distributor has enough supplies in the event of a fuel crisis or a truck drivers' strike.

For these reasons, electricity turns out to be a more efficient system in the long run, if we consider true efficiency not only in terms of cost per kilowatt hour or cost per gallon of oil, but also in terms of maintenance, repair bills, replacement parts, initial investment, and just plain convenience.

Finally, there are many jobs that can be performed only by electricity. Just think of all the things around the house—lights, washing machine, refrigerator, toaster, telephone, television and dozens more—that cannot run on coal, oil or gas. This, of course is also true for commercial buildings which employ elevators, escalators, electric typewriters and calculators. Electricity is, by far, the most versatile of all energy forms.

The availability of electricity affects many aspects of our lives—communications, for example. Telephones, picturephones, facsimilies (transmitted printed materials), computers and so forth allow us to have business meetings, produce newspapers, magazines and books, transfer large amounts of information and do our banking or shopping without traveling, which, in turn, reduces our consumption of fuel and the wear and tear on our highway systems.

The great advantages of an all-electric energy society are rapidly becoming apparent to many leaders in science, industry and economics. If our homes, offices and commercial buildings operated exclusively on electricity, no fossil fuels would be needed at these locations. All that would be required would be the delivery of electricity. The full burden of consuming fuel to generate energy would rest with the electrical power utilities. There are many advantages to this setup. It would maximize the great flexibility utilities enjoy regarding the fuels that can be used to produce energy. Because of their huge size, utilities can afford to store large quantities of fuel such as oil or coal, and therefore can obtain a better price on some types of fuel. The fact that they can burn coal is an advantage in itself; the consumption of oil is reduced, along with reliance on oil imports. Furthermore, if the natural gas and oil shortage really got bad, utilities would be better equipped to face the situation.

Electrical power plants could be built next to coal mines for maximum efficiency. Coal gasification plants located at mine sites could fuel electrical power plants. Our power utilities could even burn garbage and other solid-waste materials that we have in abundance. Add to this the other methods of generating electricity—atomic power, solar energy, geothermal, wind and tidal power—and it becomes clear that electrical power utilities have enormous flexibility and thus are not at the mercy of foreign control, labor strikes and material shortages.

There are also benefits to reap in the area of pollution control. Under our present energy system the generation plants, such as home heating, hot-water systems and industrial and commercial heating plants, are scattered by the millions throughout the country. And each one involves problems of fuel delivery, maintenance, efficient operation and pollution. It is very difficult for people in private homes and small businesses to reduce the amount of pollution emanating from their energy systems. In many cases maintenance is not performed on a routine basis and the system is not operating at peak efficiency. The result is a relatively high level of pollution (not to mention waste of fuel). It is true that there are many devices designed to reduce pollution, but these are costly and require considerable servicing and maintenance. In most cases they are too expensive for individuals to buy and have serviced. If millions of antipollution devices were installed, there would not be enough federal or municipal inspectors to ensure that they were operating at the proper level of efficiency.

The entire picture changes with large utilities. Not only can they afford large, sophisticated antipollution systems, but they also have the trained personnel to make sure that systems work properly. And because of the reasonably small number of locations, there would be enough inspectors to ensure that these antipollution systems were maintained at peak performance.

Because of the demonstrated superiority of electricity, it is anticipated that its consumption in the United States will increase substantially in the next decades. Most observers believe that electricity use per capita will quadruple by the end of the century, going from 7500 kwh in 1970 to 35,900 kwh in the year 2000.

For the past three decades there has been a natural tendency to turn to electricity as an energy source. We have new machinery operating on electricity in industry, the advent of electronics, the use of computers, increased use of radio, hi-fi, television and above all, a host of new home electrical appliances, from can openers to table-top radar ovens. More appliances continue to be added such as electric tooth brushes and vibrating beds. They add up to the need for a lot more electricity. In 1960 only 12.8 percent of American homes had air conditioners; by 1970 this number jumped to 40.6 percent. What this all means is that more and more energy will be in the form of electricity. For example, in 1971, about 25 percent of U.S. energy was used to produce electricity. By 1980 almost one third of our energy will be needed for electricity. And by 2000 approximately 42 percent of the energy in the United States will go for generating electrical power.

This trend is quite logical. After all, electric power production provides for the most effective use of coal and for the conversion of nuclear fuel to useful energy. John W. Simpson, president of the Westinghouse Electric Corporation, has argued that "the shift to an electric energy economy must become the keystone of our national energy policy," an opinion that is being voiced by many leaders in industry. But critics argue differently.

Critics claim that electricity is an inefficient form of energy, that the system of generating electrical power is inefficient in itself and that additional losses are incurred in bringing the power to the point of use. As pointed out earlier, the true cost of energy includes many hidden factors, but even if these factors are ignored, electricity is still very competitive in terms of efficiency. First of all, in resistance heating electricity is 100 percent efficient; it is converted completely into heat. And generating systems are continually improved. In 1900 electrical power plants were only 5 percent efficient; by 1973 they were 39 percent efficient; and by 1985 combined-cycle systems could attain an efficiency of 50 percent or higher.

Ways to use electricity more efficiently are being developed. The heat pump is a good example. It is a device that works somewhat like a refrigerator in reverse. What it does is to extract heat from the air outside a home or commercial building and release it inside. A heat pump can be used to warm a building in the winter and, by reversing the system, to cool it in the summer. An electrical heating system using a heat pump could be even more efficient, and cleaner, than a gas system, presently one of the most efficient methods. Such a system is now being used in the all-electric Hanford School in Richland, Washington, and it is proving to be a real money-saver in fuel bills.

There are other ways in which electricity can help us solve some of the more critical problems of the energy crisis. In transportation it could eliminate the need for gasoline. Just over the horizon is a totally new mode of transportation—the electric car, the electric bus and the electric truck. Since 25 percent of all the energy we use in the United States is for transportation, mostly surface transportation (buses, trucks, cars), consider how much petroleum could be saved if we used electric vehicles. There are about 200 million vehicles presently on the nation's roads, and approximately 95 percent of the fuel for these vehicles is supplied by petroleum, so we are talking about reducing our consumption of oil by millions of barrels per day.

The opponents of the electric car maintain that it cannot meet the requirements of the average motorist. Their argument goes like this. A tank of gasoline contains about 400 times more energy than does a battery of equal weight. This means that an electric car is extremely limited in range, unless it carries thousands of pounds of batteries. Furthermore, whereas an auto gas tank can be filled in a few minutes, it takes about six hours to recharge the batteries of an electric car. The answer to this argument is that the energy difference between the internal combustion car and the electric car is not really 400 to 1.

First of all, the best internal combustion engine operates at only 20 percent efficiency. That narrows the theoretical gap to 80 to 1, still a formidable difference. But since a car cruising at 50 mph uses only 25 percent of the fuel's energy, the difference between the gasoline car and the electric car is really only 20 to 1. And a great deal of research is presently going on to develope a practical electric car. Electrochemists are working hard to come up with more advanced types of batteries (the heart of an electric car). A battery is a device to store energy in the same way that a gasoline tank stores gasoline. If a battery can be made to store a lot more energy, the operating range of the electric car can be extended from its present 75 miles or so to as much as 200 miles. Scientists are also working on a number of new approaches. Perfection of the zinc-air battery, the aluminum-chloride battery, the lithium-sulphur or the sodium-sulphur battery may go a long way in solving the electricity storage problem. It may be possible to build a battery that can be recharged rapidly and take many recharges, at least 1000.

Actually, the electric car is already here. The Westinghouse Electric Corporation is presently operating a number of test vehicles on mail delivery routes. In a typical test run the electric mail truck travels an average of 20 miles, makes 318 starts and stops and consumes only one-tenth of the energy used by a gasoline-operated truck. But the United States is far behind in the use of electric cars. In the small country of Switzerland about 2000 electric vehicles are in full operation. Japan and Great Britain are also introducing large numbers of electric cars into their transportation systems. The Toyota Automobile Company has introduced an electric five-passenger car that can travel at 63 mph and has a range of 95 miles. However, the great market value of electric cars is beginning to be recognized in our country and a number of U.S. companies are building such vehicles. Vanguard, Inc., of Augusta, Georgia, has a car whose fuel cost is equivalent to that of a car that gets 60 miles per gallon of gas, and it has a range of 100 miles. Other companies such as Autodynamica, Inc., of Marblehead, Massachusetts, and Battronic Truck Corporation of Bogertown, Pennsylvania, are actively participating in the race to capture the potentially huge electric car market.

For many reasons, the electric car makes a lot of sense. The equivalent mileage (in fuel cost) for a small electric car can exceed 60 miles per gallon. Then there is durability. Except for wheels, electric motors and a few other simple parts, there are practically no moving parts in an electric car. An electric car built with an aluminum body can last up to 25 years. From the standpoint of maintenance, the electric car has it all over the internal combustion engine car. There is no fuel, water pump, starter motor or muffler to replace. There is no radiator to fill with antifreeze and no spark plugs or ignition points to adjust or replace. No periodic tuneups, no oil changes. No oil or air filters to buy. No freezing in winter or overheating in summer.

According to the president of the Westinghouse Corporation, "With a good job of product development and marketing, it should be possible to have 5 million electric vehicles in service in the United States by 1985 and 100 million by the year 2000." But the market is not limited to cars for private use. Electric vehicles will include police cars, delivery trucks of all kinds and buses for mass transportation. With electricity, we can develop ground transportation systems that will enable us to travel at 500 miles per hour, reducing the need for short-distance air travel.

Once we learn how to generate electricity economically, another problem remains to be solved before we can take full advantage of this best form of energy—the problem of storing electrical energy. This is necessary for three reasons. First, improved methods of storage will allow us to use electric transportation profitably. Second, consumption of electricity by the public and by industry is not constant. There are times of peak or maximum use and times of minimum use. A perfected storage system could make the generation of electricity during peak-load conditions much more economical. Finally, the ability to store large amounts of electrical energy would allow us to use other forms of energy such as sun, wind and tidal power.

How does one go about storing this elusive force? So far, only nature has been able to accomplish this feat, and it has taken her a long time. Coal, oil and gas were created over a period of hundreds of millions of years. Now man is hoping to solve the problem of storing large amounts of energy in the next decade or two.

Actually, we have already taken some important steps. About 125 years ago scientists were using electrical batteries to store electrical energy. Today this device is a common household item; transistor radios, portable hi-fis, flashlights, toys and dozens of other things work on electrical energy-storing batteries. But today's batteries cannot store large amounts of energy; they are not powerful enough and they do not last long enough. For example, the average type C or D battery loses a great deal of its power if left unused for a year or so and cannot be recharged easily.

The most common types of batteries today are the carbon and zinc device used for flashlights, portable radios and the like, and the lead-acid battery found in most cars. However, entire new families of batteries, such as nickel-cadmium, lithium-sulphur, sodium-sulphur, lithium-chlorine, zinc-chlorine and one that goes by the fancy name of poly-carbonmonoflouride, are in various stages of development. Lithium batteries have been designed to stand idle on the shelf for five years without losing their power. They can operate in arctic temperatures ($-65°$ F.) and in blistering heat ($165°$

F.). The lithium battery is smaller than a conventional flashlight battery but it carries a greater punch. At room temperature a lithium battery with a 1-amp drain (the amount of electricity flowing out of the battery) is four times more powerful than the ultramodern mercury-zinc battery, five times more powerful than the alkaline-manganese battery and seven times stronger than the magnesium battery. In fact, the little lithium battery is equal in power to 30 conventional carbon-zinc D type batteries. The National Aeronautics and Space Administration has designed a battery that will provide electrical power to satellites and deep-space probes for seven years. It uses nickel-cadmium and weighs about 170 pounds.

The ability to store a lot of power in a small package is critical for electric transportation. An efficient and practical automobile could not carry thousands of pounds of batteries. Scientists have estimated that to be efficient in a car, a battery should contain about 200 watts per kilogram of weight. The General Motors research labs have developed a lithium-chlorine battery that contains 624 watts per kilogram. But there are problems with the newer batteries. For one, lithium is a very violent chemical and must be stored in a watertight container. For another, the battery's cost, which could represent the major part of an electric car price, is still prohibitive. Then there is the problem of recharging. Most batteries require many hours to recharge and cannot be recharged too often or they will break down. In order to be feasible for use in an electric car or as a storage device for utilities, a battery must be able to stand up to thousands of recharges.

Scientists are getting close to solving these problems. General Electric, for instance, has designed a battery that can be fully charged in 15 minutes without damage, and some advanced batteries are designed to last 20 years. So it won't be very long before batteries will be powering millions of cars throughout the United States and the rest of the world. But it is still debatable whether batteries are the best means of storing large amounts of electricity by power companies. Scientists are looking at a number of other methods.

One is called pumped storage. This system pumps a large amount of water to store energy. For an illustration of how it works, see Figure 6. During low-demand hours, usually at night, a large pump driven by an electrical motor pumps water from a river or lake, up a large pipe and into a man-made lake or reservoir. When there is a large demand for electricity, a special control gate is opened which allows the water in the elevated reservoir to descend like a waterfall and drive a water turbine (hydroturbine). The hydroturbine is connected to an electrical generator, which provides the additional power required for the peak demand. It may sound crazy to use electricity to pump water up a hill so that the water can fall

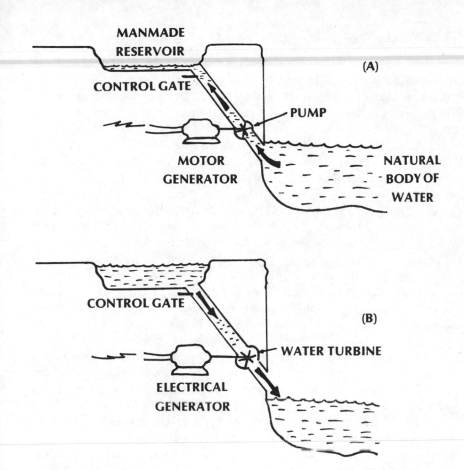

MANMADE
RESERVOIR

(A)

CONTROL GATE

PUMP

MOTOR
GENERATOR

NATURAL
BODY OF
WATER

CONTROL GATE

(B)

WATER TURBINE

ELECTRICAL
GENERATOR

FIGURE 6. Pumped-water storage system stores energy in the form of water in the elevated reservoir (A). The water then falls to drive a turbine, which runs an electrical generator (B).

downhill to generate electricity. But using electricity to pump water during low-demand hours, and then later, during peak-demand hours, using that water power to generate electricity, saves a lot of money. That it is cheaper than using inefficient small electrical generating units such as fossil-fueled or gas turbines to supply peaking power has shown to be true in actual operation. Pumped-water storage systems are not new—in the 1930s a 32-MW system was operating in Connecticut and today there are many others including a 1900-MW system in Detroit—but they are not without problems.

This type of system demands a great deal of land and an area of high elevation above a natural body of water. There is a limited number of such sites in the country. Even where the sites are available, local opposition is often considerable because of the interference with aquatic life and the damage to the environment. In a study made by the Harza Engineering Company of Chicago 100 possible pumped-water storage sites were surveyed. It was found that for 50 percent of these there would be insurmountable conflicts with cultural, recreational and industrial uses. Another 10 percent were eliminated for geological reasons, and environmental conditions ruled out another 15 percent. Only 25 out of 100 sites promised any chance of success.

Furthermore, pumped storage sites are often located far from the areas they serve, so long-distance lines must be constructed. This entails additional energy loss (in the power lines) and high-cost of construction, not to mention the objections by environmentalists to large transmission towers that spoil the natural beauty of the landscape.

Finally, according to some critics, long lead time for planning and construction, and high capital cost involve an inherent inflexibility rendering a company unable to respond to changing needs.

Another version of pumped storage may be the answer. Instead of pumping and storing water, this system would pump and store compressed air. Why air? Because a lot of electricity is produced with gas turbine generator systems, and gas turbines are like jet engines. In a jet engine air enters the front or intake (see Figure 7). The front part of the engine has a fanlike device called a compressor which packs the air tightly or compresses it. This air, which is now pressurized, enters another part of the engine called the burner chamber where fuel is sprayed and ignited. The hot burning gases that result from the combustion flow through a turbine

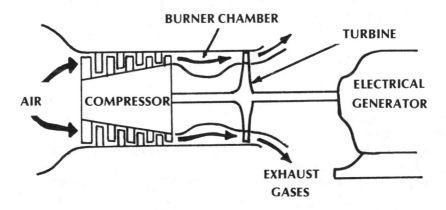

FIGURE 7. A gas turbine engine, which works like a jet engine, is used to drive an electrical generator.

wheel, causing it to rotate. The turbine then drives an electrical generator, which produces electricity.

Not all the energy of the turbine wheel is used to drive the generator. Over 65 percent of it is needed to rotate the compressor, which sucks air into the engine and compresses it. And this is the key to compressed-air energy storage. During off-hours when the demand for power is very low, the extra electricity can be used to operate a motor (one of the electrical generators actually acts as a motor). The motor is connected to the compressor portion of the gas turbine (jet engine—see Figure 8A). As it rotates the compressor, the compressor pumps large amounts of air into a large pipe connected to a large natural cavern. The air pressure in the cavern continues to build, and energy is thereby stored in the form of

FIGURE 8. Compressed-air storage systems do not require large land areas and can increase the efficiency of gas turbine engine electrical power systems.

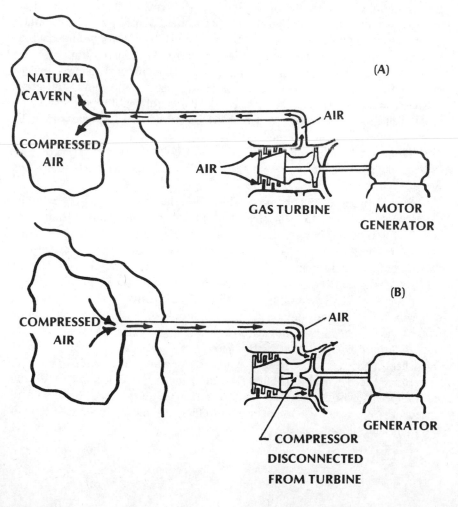

compressed air. During peak hours the compressed air is allowed to flow out of the cavern directly into the burner chamber of the gas turbine engine (Figure 8B). Fuel is injected and ignited, and the hot gases drive the turbine wheel. And since the compressor portion of the engine is disconnected from the turbine, all the energy of the turbine wheel is used to operate the electrical generator, which produces electricity. This very efficient system can cut turbine engine fuel oil consumption in half, while at the same time increasing the amount of electricity generated.

Still another way to store energy is to store air in liquid form, that is, take the air being pumped by the compressor and freeze it down to a liquid. This method would not require a large storage area such as required for compressed air, so the entire facility could be located close to the area where the electricity is being used, instead of near the closest available cavern. When the air was needed, the liquid could be heated and large volumes of air would flow into the turbine engine. In fact, the idea of storing liquid air solves one of the biggest problems of the compressed air storage system—the availability of natural sites. Since it would be much too expensive to excavate a huge cavern, the electric utility would have to be located at the cavern, which might require the construction of long-distance, high-voltage transmission lines. Of course, natural caverns are not the only things that will work; old mine shafts, unused salt mines and other excavations could be used for compressed air storage. In any event, the amount of area needed for a compressed air storage system is less than 15 percent of that needed for pumped water storage.

What needs to be done now is for a demonstration compressed air storage plant to be built so that all the bugs can be worked out of the system.

Scientists and engineers are working on other energy storage methods, some of which appear to be straight out of science fiction. One ultramodern approach has actually been with us from ancient times. It is the basic idea behind the potter's wheel, also known today as the flywheel. Present-day flywheels are usually made out of steel or cast iron. They are designed to be very heavy and bulky so that once they start spinning, they will keep going for a while. The principle is this: The flywheel has weight; so in order to move or rotate it, force has to be applied. As the force is applied to the wheel, it begins to spin (see Figure 9A). This spinning motion is a form of stored energy—that is, the force has overcome the momentum of the wheel, and once the wheel is in motion, it tends to remain in motion. The faster the flywheel is rotated, the more energy it will store. Also, the heavier the flywheel, the more energy it can store. As we shall see, the speed of a flywheel determines storage efficiency more than the weight if the right materials are used.

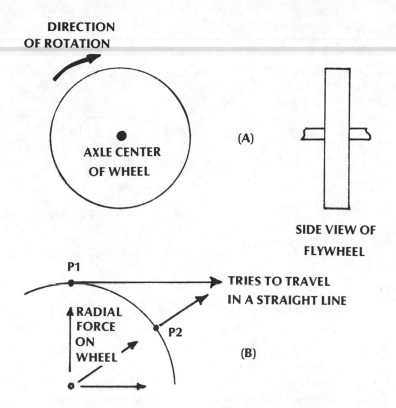

(A)

AXLE CENTER
OF WHEEL

SIDE VIEW OF
FLYWHEEL

P1

TRIES TO TRAVEL
IN A STRAIGHT LINE

RADIAL
FORCE
ON
WHEEL

P2

(B)

FIGURE 9. Flywheels in operation can store energy. Force applied to wheel causes it to rotate, thereby storing energy (A). Rotating flywheel creates forces that push outward from the center, trying to destroy the wheel (B).

One problem with flywheels is that as they rotate faster, they build up a force that tries to shatter the wheel. Any point on the wheel (P 1 in Figure 9B) tends to travel in a straight line, though the structure or shape of the wheel forces it (P 2) to move around the curve of the wheel. Still, that point is trying to pull away from the wheel, and this is true for all points on the flywheel. The result is forces that are radiating outward from the center of the wheel. These radial or centrifugal forces place a great amount of stress on the wheel. If the wheel spins too fast, it will shatter into pieces that fly in all directions. Today's flywheels are not strong enough to store large amounts of energy. This has been a serious limitation until now, but modern science is coming up with some good solutions.

One way of partially getting around the problem of centrifugal force is to construct the flywheel with a different shape. Instead of being in the shape of a disc, as seen in Figure 9A (side view), the flywheel can be made

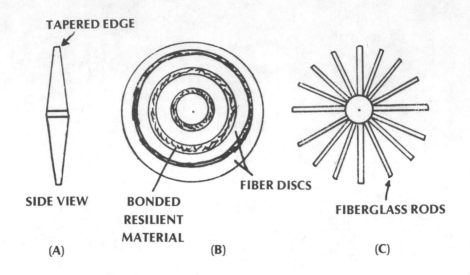

TAPERED EDGE

SIDE VIEW

BONDED
RESILIENT
MATERIAL

(A)

FIBER DISCS

(B)

FIBERGLASS RODS

(C)

FIGURE 10. High-speed flywheel designs: (A) tapered; (B) concentric rim; (C) superflywheel.

very thick at the center and tapered toward the circumference (Figure 10A). Since the greatest part of the force on a flywheel increases at the outer portions, concentrating most of the weight at the center of the wheel will reduce the centrifugal force on the wheel. This principle is applied to all high-speed rotating discs such as jet engine turbines.

Another way to increase the strength and energy-storing capability of a flywheel is through the use of new, superstrong, lightweight materials. One family of these materials is the various types of glass fibers. Glass fibers are extremely light and very strong—about four to six times lighter than steel with greater tensile strength. And they are even cheaper. For example, E glass can store four times as much energy per unit weight as can high-strength steel, yet it costs about 85 percent less. Other fiber materials such as Du Pont's PRD-49 (Keylar) and products being developed by Monsanto can store seven times as much energy in a flywheel as can modern alloy steel. Fused silica fibers produce flywheels that can store up to 15 times more energy per unit weight than can flywheels made from the best alloy steel.

Another approach is to use glass fibers to make a flywheel consisting of a series of discs, one inside the other (see Figure 10B). Each disc would be separated by a band of bonded resilient material, which would hold the discs together. Such a design would make efficient use of the flywheel's volume. However, this design has one problem. At very high speeds the discs would try to separate from each other, since the force on the outside discs would be greater than the force on the inside discs. So scientists have come up with still another design. It is called the superflywheel.

The superflywheel is also made of exotic materials such as boron filaments, fiberglass and carbon fibers. But instead of a full disc, the flywheel consists of a series of rods sticking out from the edge of a small disc (see Figure 10C). This design reduces the overall weight of the flywheel while maintaining its strength. Superflywheels could attain very high rotational speeds, thereby storing large amounts of energy. But how do they tie into the energy crisis?

First of all, flywheels could be very valuable as energy storage devices in the electrical power industry. They are quiet, they cause no pollution and they are about 95 percent efficient. Furthermore, flywheels require only a small fraction of the land that a pumped-water storage facility or even a pressurized-air storage system does. For example, for 10,000 kwh of stored energy, a pumped-water storage system typically requires between two and four acres of land for a reservoir. A flywheel unit storing the same amount of energy would need a space only 20 feet square. Then there is cost. As production increases, the cost of materials for flywheels will decrease—unlike the cost of land, which keeps rising. The flywheel would be used to generate extra power during high electricity demand periods. When demand is low, the extra electricity would be used to operate motors that would rotate flywheels at great speeds. A motor-generator would be built right into the flywheel structure (see Figure 11). When demand for

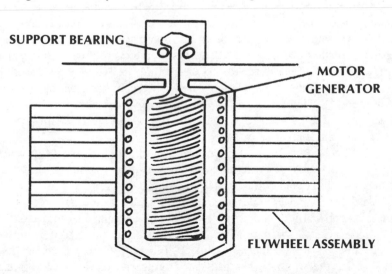

FIGURE 11. Simplified illustration of a peaking-power energy storage assembly. The flywheel would measure 15 feet in diameter and weigh about 200 tons.

electricity increases, with the pressing of a few control buttons (with an automatic control system), the flywheel assembly would become a self-powered electrical generator. A typical unit might store from 10,000 to 20,000 kwh. These units would measure 15 feet in diameter and weigh about 200 tons. An urban utility system might use 100 or more of these flywheel energy storage units. Some would be located in the main plant, while others would be installed in substations around the city. They would produce all the power needed without adding an ounce of pollution to our air, land or water.

But there are further uses for the flywheel. It could also be valuable as an energy storage device in other types of power-generating systems such as tidal wind, and even solar power. These subjects will be discussed more fully in subsequent chapters. Here the discussion will be confined to an area where superflywheels can do a great job in saving fossil fuels for the United States—transportation.

In San Francisco there are two trolly-buses operating daily. They make their rounds, stopping and going like conventional vehicles, except for one thing—they require no gasoline. They are operating on flywheel power. The trolly-buses were built by the Lockheed Missile and Space Company, and they might be the beginning of a new trend in transportation. For example, your family car could run very well on flywheel power. Such a car would be extremely quiet, highly efficient and pollution-free. Flywheel-powered cars even have advantages over battery-operated electric cars. They would be lighter. A 130-pound flywheel would store about the same amount of energy as would 2000 pounds of lead-acid batteries. This same flywheel could power a car on a 200-mile trip without the need of re-energizing. Furthermore, while a lead-acid battery requires hours to re-charge, a flywheel could be recharged or spun up in five minutes. A battery suffers a certain amount of damage when it is recharged, which is why it has a limited life. There is practically no limit to the number of times that a flywheel can be reenergized (eventually there is slight wear). And it cannot be beaten for efficiency.

A flywheel-powered system uses about five times less (equivalent) energy per mile than does an internal combustion engine. And it is an extremely simple system with very few moving parts. Figure 12 illustrates a simplified flywheel-powered car. A motor generator would be built into the flywheel assembly. Initially, the motor would be energized from an external source of electricity—perhaps your local service station, which would sell electricity instead of gasoline. (Instead of "Fill it up," one would say "Spin it up.") Once the flywheel was at its operating speed, it would rotate the generator, which would produce electricity. This electricity would be routed to four electric motors, one at each wheel. The motors would be the

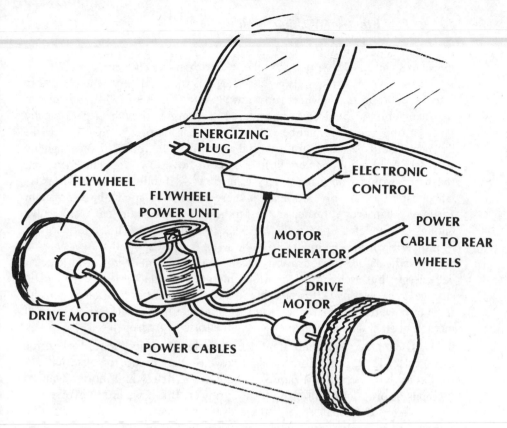

FIGURE 12. Cars powered with superflywheel units are silent and more efficient than gasoline-powered cars, and generate absolutely no pollution.

driving force for the car. It is really a very simple system. And it has a couple of other advantages. During acceleration an automobile requires a lot of extra energy. With a combustion engine this energy is in the form of extra fuel supplied by the carburator. In a battery-operated electric car acceleration is a problem. The extra energy needed is in the form of electrical current, and batteries cannot supply the huge amounts of current required for fast acceleration without suffering damage. But the flywheel system can supply this surge of electricity with no problem. Finally, a flywheel car would use a lot of the energy normally wasted in a conventional car. Each time a car is braked to a slower speed or to a stop, energy is expended at the brakes in the form of heat. In an electric car the motors that drive the wheels also act as generators. When the car is braked, the force of momentum causes the four motor-generators to rotate, thereby generating electricity. This electricity is stored in batteries, or used to drive the motor generator in the flywheel assembly, spinning up the flywheel. It is this conservation of energy that makes electric cars so efficient.

The flywheel car sounds great so far, but what happens when one is in a traffic jam and the flywheel runs down? Actually, there is little chance of that happening if the driver remembers to reenergize the wheel periodically (just as one has to remember to gas up a car before a trip). Since the flywheel would be housed in a partial vacuum and supported on magnetic bearings, there would be very little friction to slow down the wheel. In fact, the wheel would rotate for a very long time—a family could drive a flywheel car to an airport and leave it there while they spent the summer in Europe. When they returned in the fall, the flywheel in the car would have enough energy left to power the car back to their home. The rundown time of a superflywheel is estimated to be between 6 and 12 months.

Flywheels are demonstrably excellent mechanisms for storing electrical energy, but there are still other methods available to us. Magnetic storage is one of them. It is possible to store huge amounts of energy in the form of powerful magnetic fields in superconductor magnets. Another good method of storing electrical energy is in the form of hydrogen.

As we have seen, there is no lack of scientific approaches for storing excess electricity. And storage is the key to efficient use of electricity as well as to the practical utilization of some of the newer and more advanced methods of producing this clean form of power. But if we are to evolve into a completely electrical society, there are other aspects of electricity that must be considered. Transmission and distribution of power, for example.

In the 1930s United States power utilities had electrical generators capable of producing about 100 million VA (volt-amperes) or 100 MW. By 1970 their large generators had a capacity 15 times greater, or 1500 MW. In 1974 total electrical generating capacity in the United States was 340,000 MW, and it is predicted that by 1990 the total will be an astounding 1,300,000 MW. Individual power plants capable of generating 11,000 MW will be in operation. The great challenge facing scientists and engineers is how to control and distribute all this power.

The power has to travel along thousands of miles of high-voltage overhead and underground cables. It must be transferred from hundreds of main generating plants to thousands of substations, from which it will be further distributed. Complex networks have grown up, such as the Northeast "power grid." A power grid is the joining or integrating of many individual electrical power generating systems. The integration is done first of all to save money. When demand is very high in one part of the grid, the affected power company can buy electricity from a member utility whose power demand may be low. Electricity can be produced at a lower cost because it comes from a large main generator, which is much more efficient than smaller peaking generating units. The second reason for integration is that one part of the grid may be short of power. To prevent a brownout (a reduction in voltage), the affected utility can buy the needed power from another part of the grid. Finally, the grid system adds an element of dependability. Each part of the system covers the other parts. If one electrical generating system fails, generators from the other systems can supply electricity to that system until the malfunctioning generators are repaired. So power grid systems are extremely important in an electrically oriented society. But they are extremely complex and can be controlled only by huge computers.

Providing enough high-voltage lines or power transmission lines for the anticipated growth in demand is a real challenge. The biggest part of this challenge is what is referred to as line resistance losses. About 10 percent of the electricity generated is lost as it travels through the power cables to the area where it is being used. Electricity is the flow of many electrons through a wire or conductor (see Figure 13). However, copper, silver, aluminum and other conductors naturally resist the flow of electrons. This quality of electrical resistance results from the collision of the moving electrons with each other or their capture by other atoms. In either case the electrons lose much of their energy in the form of heat. That's why a wire gets hot when a lot of current is flowing through it. In fact, if too much electricity is allowed to flow through a wire, the wire can melt from the intense heat. For this reason, and because of the loss of energy, engineers must design bigger and better systems to carry the huge amount of electrical power that will be needed in the next two decades.

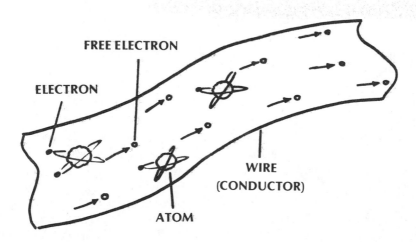

FIGURE 13. Electricity is the flow of free electrons through a wire or conductor.

Even more of a challenge will face these engineers if we decide to convert to a totally electric society. Our present power lines are now carrying their maximum capacity. There are overhead lines that are presently operating at 750,000 volts (house voltage is 120 volts). If the demand for electricity triples or quadruples, as expected, we are going to require a great number of additional power lines. But already there are thousands of miles of overhead cables and thousands of giant electrical towers throughout the countryside. Many people are opposed to adding more of these overhead cagelike structures which destroy the natural beauty of the land. Furthermore, because of high land values, the cost of right-of-way for the towers can be very expensive. Of course, the cables could be placed underground, but besides being even more expensive, underground cables are not feasible in most big cities where the area below street level is already jampacked with telephone, steam, water and sewerage, gas and existing electrical power lines. Therefore size versus power-carrying capability is important. The more power that must be handled, the bigger the size of the cable needed. It would be extremely difficult to meet the electricity demands of the 1980s and 1990s using conventional electrical power cables. Fortunately, we may not have to depend on them. A relatively new science, cryogenics, is changing the whole picture.

In the early years of this century a scientist named Kamerlingh-Onnes studied how metals react under very cold temperatures, close to absolute zero ($-459.7°$ F.). Nothing unusual happened when he worked with platinum and gold, but when he turned to mercury, something completely unexpected occurred. Onnes found that when mercury is cooled down to about $-452°$ F. it enters a new state of matter in which there is virtually no electrical resistance. Onnes called this a superconducting state. The incredible implication was a wire in which electrical current could flow with no resistance. By 1933, 17 other elements were found to have superconducting states at different temperatures. It was also discovered that mixtures of metals or alloys exhibited this characteristic, and by 1933, 53 of these combinations were identified. What do these discoveries mean to an electricity-oriented society?

For power transmission, they mean a great deal. A superconducting cable can carry 20 times more electricity than a conventional type. Such cables would take up a lot less space. But while it is easy to describe such a cable, it is hard to devise one. A cable remains in its superconducting state only so long as it is kept at its critical temperature, and in the case of mercury this is $-452°$ F. or $4.2°$ K. (Kelvin). It takes a lot of refrigeration to keep something that cold. Luckily, other metals and alloys have higher critical temperatures—lead, $7.2°$ K.; niobium, $10.5°$ K.; and the so-called hot superconductors discovered in the early 1970s such as the alloy made from niobium, aluminum and germanium, 20.8° K., and an even hotter superconductor discovered by John Gavaler of the Westinghouse Corporation, 22.3° K. Recently the record was broken by a group of scientists at Bell Laboratories, who came up with a superconductor material that operated at 23.2° K. It may not sound like it, but there is a great difference between keeping material at 23.2° K. and keeping it at 4.2° K. It is also a lot less expensive. Many scientists are confident that they can develop alloys with critical temperatures as high as 60° K. The ideal discovery, of course, would be an alloy that remains in the superconductor state at room temperature.

It is probable that developments over the next three or four years will make this technology practical. As seen in Figure 14, in a power transmission system a superconductor cable will be housed inside a tube. Liquid helium or hydrogen (depending on the critical temperature of the material used) will be pumped into the space between the cable and the outer jacket. The cable will be kept at a cryogenic (supercool) temperature by the liquid helium or hydrogen as it carries large amounts of electrical power (billions of volt-amps). At certain distances along the transmission line route refrigeration facilities will be located. Since the flow of electricity creates a small amount of heat even in a superconducting cable, this heat will accumulate and eventually raise the temperature of the liquid in the

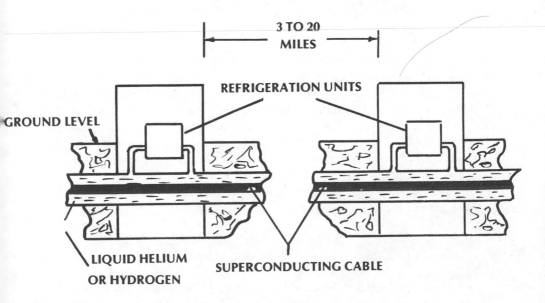

FIGURE 14. Cryogenically cooled superconducting electrical cables can carry 20 times the power of conventional cables.

jacket. If the temperature rises above the critical point of the superconducting material, the cable will "go normal," and with all that current flowing, together with the normal resistance of the line, a great amount of heat will be generated. To prevent this from happening, refrigeration units will extract all the heat from the cryogenic liquid and keep it at the required temperature. The big challenge is not building the superconducting cable network, but constructing a dependable refrigeration system.

Superconductors will also allow huge power generation units to be reduced to less than half their size. The generators presently in operation are of low capacity. Eventually large units capable of generating 100 billion watts of power will be built, which could be used as the main generators in power stations. Superconducting generators have higher efficiencies than conventional units: in the higher-power models energy losses will be reduced by 50 percent. In the long run superconducting generators will be cheaper to build.

As all these developments take place, the superconductor technology will be used in other areas to fight the energy crisis. One major area is mass transportation, specifically high-speed rail travel. Conventional trains are limited to about 150 mph. But because of road bed, track conditions and wheel wear, even trains designed for this speed, such as the high-speed line out of Tokyo, Japan, rarely travel faster than 120 mph. Yet with superconductor technology there is no reason why we could not travel at speeds of 300 or 400 or more mph along a ground-based system.

Using superconducting materials, very powerful magnets can be built, magnets capable of lifting a tremendous amount of weight—a train, for example. The train illustrated in Figure 15 will attain speeds as high as 500 miles per hour. Mounted on a large steel guide rail, it will have two large superconducting magnets, one on each side. Above each magnet will be a chamber containing cryogenic liquid to maintain the magnets in a superconducting state. In the center of the train, suspended above and around the guide rail, will be a special type of motor called a linear induction motor, which works on the basis of magnetic field repulsion and attraction. In other words, when the north poles of two magnets are placed close to each other, they push or repel each other. If opposite poles (a north and south pole) are brought near each other (Figure 16), they will pull toward each other. In operation, the superconducting side magnets will suspend the train above the guideway a fraction of an inch. Using magnetic fields, the linear induction motor will pull the train along a large steel guide rail. Because the train will be suspended in air, there will be no serious friction problem and the train will race smoothly and quietly along the guideway at very high speeds. When it slows down or comes to a stop, the operating magnetic fields will be reduced and the train will settle on a set of conven-

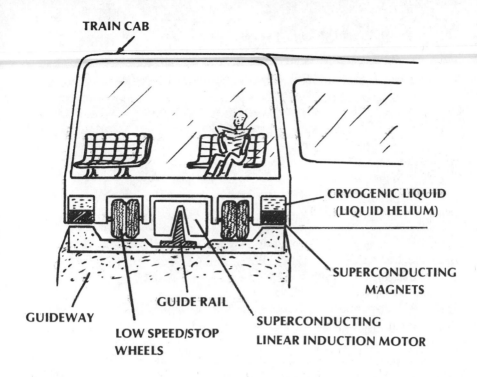

TRAIN CAB

**CRYOGENIC LIQUID
(LIQUID HELIUM)**

**SUPERCONDUCTING
MAGNETS**

GUIDE RAIL

GUIDEWAY

**LOW SPEED/STOP
WHEELS**

**SUPERCONDUCTING
LINEAR INDUCTION MOTOR**

FIGURE 15. Superconductor-powered levitation trains can whiz along at 500 mph.

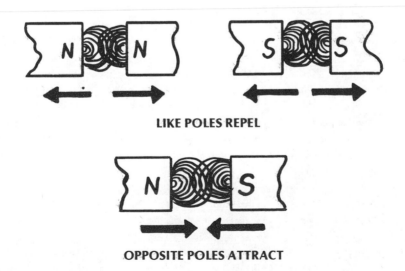

LIKE POLES REPEL

OPPOSITE POLES ATTRACT

FIGURE 16. North poles of two magnets will repel each other. Opposite poles will attract each other.

tional rubber wheels. Such a train would be much more efficient than a plane. It would use a lot less fuel, and tremendously reduce present levels of pollution.

Besides saving energy, superconductor technology would permit the storage of vast amounts of energy. Because of their zero resistance, small amounts of current in superconductor coils can induce extremely high magnetic fields in an iron core. Magnetic fields contain energy. So if a huge superconducting magnet were to be built, electrical energy in the form of a powerful magnetic field could be stored in its core. As was explained before, a superconducting wire offers no resistance to the flow of electrical current: the current continues to flow, without the need for additional energy, for a long time. How long? After 10,000 years about half of the original electrical current will still be flowing in the loop of wire. The energy in a superconductor storage system could be extracted for later use with an efficiency of 95 percent. Obviously, this is an excellent storage method: it does not require huge amounts of land, it creates no pollution and it has no moving parts. Of course, like everything else, it does have a drawback.

Magnetic energy storage systems are not economically feasible on a small scale. Unlike the flywheel, which can be located at many points in an electrical distribution system, a magnetic storage system would have to be large enough to store between 1 billion and 10 billion watt-hours of power, or enough power to keep a 100-watt light bulb burning for 100 million hours. A typical system would be made up of a doughnut-shaped structure about 170 feet in diameter and 170 feet high. Despite these problems, magnetic energy storage is more than just an imaginative theory. The University of Wisconsin and the Los Alamos Scientific Laboratory are both doing a great deal of research and development in this area and are planning to build demonstration magnetic energy storage systems within the next five years.

Superconductor technology is one possible answer to the challenge of large-capacity electrical energy transmission. There is another even more advanced and exciting method of solving this problem. Imagine a large electrical power station producing huge amounts of energy, then beaming it up to a satellite suspended in space. The satellite, which is a huge reflector, transmits the electrical energy back down to the earth (see Figure 17), where it is to be used. This is not science fiction. The theories have all been worked out and such a system is very possible. Basically, the electrical power generated at the utility would be converted into another form of energy—high-frequency (microwave) electricity. This microwave energy might be over a billion cycles per second. (House currents are 60 cycles [hertz] per second.) Using a large transmitting antenna, the microwave

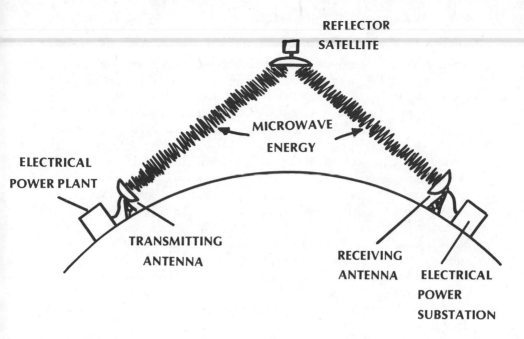

REFLECTOR
SATELLITE

MICROWAVE
ENERGY

ELECTRICAL
POWER PLANT

TRANSMITTING
ANTENNA

RECEIVING
ANTENNA

ELECTRICAL
POWER
SUBSTATION

FIGURE 17. Electricity could be transmitted over long distances through the use of microwave energy and satellite energy reflectors.

energy would be beamed into space to a satellite, which would have a large reflecting surface to bounce or redirect the electrical energy back down to the earth at a location considerably distant from the power station. This location would have a large receiving antenna to collect the energy beamed from the satellite. The microwave energy would be changed back to normal house current at a substation and distributed to the local area for general consumption.

There are many advantages to such a system. Since there would be no need for thousands of huge electrical towers on the ground, esthetic pollution would be reduced. More importantly, it would be possible to locate large generating plants far away from populated areas, which would reduce local pollution and also allow utilities to build power plants on low-cost land. According to Dr. Krafft Ehricke, executive adviser to the North American Rockwell Space Division, a satellite energy relay system would work well for solar power plants located in isolated desert or ocean areas. The large transmission distance between the plant and the area of consumption could easily be handled by such a system.

Electrical energy transmission and storage certainly are not stumbling blocks in the way of the totally electric society. Considering the availability of such methods as superconducting lines, microwave transmission, pumped-water storage, compressed-air storage, superflywheels, chemical (hydrogen) storage and magnetic storage, we can easily meet future demands. What is required is a more intensive research program to determine which of these methods will be the most practical. Most likely, all will be practical to some extent. But there is little time to waste. The trend toward electricity is accelerating because electricity answers our energy needs in the most effective way. With full use of electricity we will enter an era of clean, quiet energy.

But this will not be cheap. It has been estimated that building all the new electrical production capacity required over the next 30 years will take 19 billion engineering and construction man-hours, exclusive of the man-hours needed for the development, design and manufacture of generating units, boilers and associated machinery, and for the transmission and distribution networks. These estimates refer only to the task of keeping up with the normal increase in electrical power consumption. If we were to accelerate more rapidly toward an all-electric society, the numbers would double or even triple. Can it be done?

The answer, by necessity, must be yes, it can be done. We really do not have much of a choice. Either we attack the problem with all our energies and complete the task, or we procrastinate for a decade or so and then face far more distasteful consequences. There are many reasons why we must evolve to a totally electric society as rapidly as possible: we are rapidly running out of fossil fuels; fossil materials should not be wasted as fuels because they are needed for thousands of products; pollution is affecting us all adversely; and international politics are uncertain and dangerous. The question is not *whether* to turn to electricity, but which are the *best methods* of generating the vast amounts of electrical power that will be required by the United States. It is to this question that the following chapters are addressed.

3

Tradeoffs for Today's Electricity

I t is clear to most scientists that electricity is the best form of energy. The critical factor is how we generate it. It would be great to have this pollution-free energy, but what are the tradeoffs involved in producing it?

What good is pollution-free electricity if the plants that generate it produce huge amounts of pollution?

What is the benefit of using this petroleum substitute for energy in our homes if our electrical utilities have to consume vast amounts of petroleum to generate this power?

What is the sense of converting to a form of energy that reduces the danger of environmental degradation if the method of producing that energy places us in a situation far more dangerous?

What does it avail us to turn to electricity to be independent of foreign influence if the method we chose to produce the power has the reverse effect?

Producing electricity is only half the answer to the energy crisis. The other half concerns *how* we produce electricity. What methods will we use? Are these methods safe? Are they pollution-free? Do they depend on limited fuel supplies or are they based on self-renewing energy sources? Are they in tune with nature? Are they feasible and economical? These questions must be answered positively if we are to become a successful electricity-oriented society.

As we have seen, there is a definite trend toward electricity as a universal form of energy. Consequently, electric utilities are among the biggest businesses in the country. Unfortunately, wisdom, or even plain common sense, has not thus far been applied to discovering the best methods of producing electricity. Without question, there have to be drastic changes in today's methods of generating electrical power: their efficiencies have to improve; their effects on the environment have to be minimized; and their dependency on limited and very valuable fossil (fuel) materials must be discontinued.

From the beginning of the electrical industry (1882) hydroelectric power was the preferred method of generating electricity because of the vast amount of power available in the form of waterfalls. In 1903 the first steam turbine electrical generating plant began operating. But steam turbine systems could not compete with hydroelectric systems because of their limited capacity. Furthermore, since steam turbine generators consumed a great deal of fuel, they had very low efficiency. So for the first two or three decades of this century the major burden of generating large amounts of electrical power was carried by hydroelectric plants.

Today there are 1507 hydroelectric plants in the United States where electric power is produced. Unfortunately, there are few usable sites left for this method; we've thrown dams across most of our major rivers. Of course, even without new sites hydroelectric power could be increased (it presently supplies 4.1 percent of the nation's energy). Additional generators could be installed at many of the sites, which could increase hydroelectric generating capacity by 25 percent by 1980. With the advent of superconducting technology, generators capable of producing much more electricity will be available. If superconducting generators replaced conventional units, hydroelectric power production could possibly be doubled.

Hydroelectric power, however, is not without problems. Although it produces little air pollution, it has other drawbacks. One is that it often requires the flooding of large land areas, as in the case of the Hoover Dam. This may disturb local wildlife as well as alter the nature of the water on the upstream side of the dam, changes which affect the viability of local aquatic life and vegetation. Another drawback is that seismic pressures resulting from the hydrostatic pressure of the accumulated water behind the dam can cause subterranean fractures and ruptures. So although hydroelectric power is preferable to other methods of generating electricity (such as coal- and oil-fired steam turbines or atomic power), it is certainly not ideal.

Hydroelectric power is only a small fraction of total power generation; the lion's share goes to steam turbine power systems. Although early steam turbine systems could not compete with hydroelectric power, they eventually improved. After World War II large steam turbine units began operation; with the availability of cheap oil and natural gas as fuel, and with their increased efficiency (about 25 percent), steam turbine systems quickly overtook hydroelectric power. Today steam systems have reached 40 percent efficiency and provide almost 75 percent of the nation's electricity. Steam turbine systems (the basic operation is described in Chapters 1 and 2) will dominate the electrical power industry for at least the next twenty years. Even so, other methods of generating electricity will be required to meet demand before the end of the century. One method, which has been in operation since the late 1950s, is atomic or nuclear power.

When we consider all the aspects of the energy crisis, nuclear power seems a sensible solution. Our dwindling supplies of fossil fuels, the requirement of fossil fuels by other critical industries, their rising cost, the pollution problems resulting from using fossil fuels, the political implications of fuel distribution—all demand that we search for a long-range energy supply that will eliminate these problems and place the United States in an independent energy position. Nuclear power appears to have the capabilities. On December 2, 1942, the first controlled nuclear chain reaction was initiated in a crudely built carbon-pile reactor and atomic power was born. Unfortunately, so was the atomic bomb. Nine years and one world war later, on December 21, 1951, the atom was finally put to peaceful use: electric power was generated by an experimental nuclear reactor built at the Atomic Energy Commission's National Reactor Testing Station in Idaho. It produced about 200,000 watts of power. Suddenly, atomic energy became beneficent magic: the atom would solve all our energy problems, providing unlimited clean energy forever. Or so many people believed. A lot of electric utilities discarded plans for new coal-fired electric plants, believing that nuclear power would soon be available. But it was slow in coming. By 1967, 16 years after the operation of the first electric power reactor, only 14 reactors were in operation. Not only had it turned out that atomic power plants were extremely complex to build, but in the eyes of the general public this new energy dream was beginning to lose some of its appeal.

Atomic power is unquestionably the most controversial method of generating electrical energy. Many citizen groups and professional societies, such as the Union of Concerned Scientists, have fought vigorously to prevent the construction of new atomic power plants. They have presented a number of objections ranging from the effects of heat pollution to the fear of a catastrophic nuclear explosion. Government representatives, nuclear scientists and power utility managers have been equally forceful in defending nuclear energy. The general areas of controversy could be divided into the following categories:

- Heat pollution
- Limited fuel supplies
- Nuclear reactor safety
- Radiation
- Nuclear waste storage
- Theft of nuclear fuel or waste

THE HEAT IS ON

Those who oppose nuclear power claim it will add greatly to the nation's heat pollution problem. Present atomic power plants do not use the heat they produce as efficiently as do fossil-fueled plants (32 versus 40 percent). The extra heat from nuclear plants must be discarded somehow, either to a body of water or to the atmosphere. As discussed in Chapter 1, the huge amounts of heat emanating from atomic plants can have serious ecological effects.

Defenders of nuclear power claim that the extra heat need not be wasted. Professor David J. Rose of MIT suggests that heat from atomic plants (either as steam or hot water) could be very valuable for industrial or commercial applications—to process materials, run machines or heat commercial buildings and apartment houses. Another approach to the problem of excess heat is the use of artificial cooling lakes to receive the warm water coming out of a nuclear plant. It has also been suggested that a method called the outflow system would effectively handle excessive heat. In this method long pipes are installed from the power plant far out into a large body of water such as a big lake or the ocean. The warm water is pumped under high pressure so that as it leaves the pipe, it disperses far away from the shore, where its effects are greatly reduced.

Still another answer to the problem of heat-use inefficiency is a new type of nuclear power plants such as the high-temperature gas-cooled reactor (HTGR) and the breeder reactor, both of which will be discussed shortly. These forms of nuclear power use heat at about the same level of efficiency as fossil-fueled plants (40 percent).

Finally, there is a very interesting approach to the thermal pollution problem that involves placing nuclear plants about three miles off the coast of the United States. This plan also attempts to deal with other problems, such as site selection, security and reactor safety. One such plant is being built by Offshore Power Systems, which is owned by Westinghouse, the world's largest supplier of nuclear plants, and Tenneco, the nation's largest shipbuilder. This nuclear goliath will comprise two reactors, each rising 17 stories above the waterline and measuring 400 feet by 387 feet. The huge complex will include not only the reactors, but also large steam turbines, electrical generators, maintenance shops, control rooms, offices, equipment storage areas, complete living quarters for the 100 or so people needed to operate the station and even a helicopter pad. Power from the electrical generators will be transmitted to the shore by heavy underground cables. The 2300 MW of power that will flow through these cables will be enough electricity for a city of one million people.

Circling the entire nuclear power station (which will draw 32 feet of water) will be a structure of monstrous proportions. This protective wall or breakwater is required to protect the atomic plant against stormy seas and ramming by ships. It will be constructed of sand, gravel, large stones and about 69,000 dolosses (large steel structures shaped like children's jacks), each weighing from 6.5 to 42 tons. The breakwater will rise 50 feet above sea level, be 2000 feet long and be capable of withstanding winds up to 300 mph. It will be so massive and strong that it will be able to stop a giant supertanker, fully loaded, ramming it at full speed. The entire complex, including the breakwater, will cover an area of about 70 acres and cost about $1 billion.

According to Offshore Power Systems, such a nuclear power plant offers a number of advantages. Almost half the electricity consumed in the United States is used by customers located within a 200-mile-wide strip along the Atlantic, Gulf and Pacific coasts, and it is becoming increasingly difficult to find suitable sites (of 300 acres) for nuclear plants in these crowded areas. Floating atomic power plants, however, can be built on any of thousands of offshore sites along 5700 miles of coastline. Other advantages listed by the company include the cushioning or insulating effect of water against seismic shock (earthquakes), the availability of unlimited cooling water and minimum impact on the environment. If the floating nuclear plants obtain government approval, the U.S. coastline may be dotted by dozens or even hundreds of these power poseidons by the end of this century.

UNLIMITED FUEL SUPPLY

Is there enough nuclear fuel to last a long period of time? Unfortunately, the answer to this question is not a straightforward yes or no but—It depends! On what? On the price of nuclear fuel (uranium). As with natural gas and oil, if the price of uranium is too low, there will be no profit incentive for companies to explore for new sources. If the price rises, it will become economically feasible to explore for uranium ore and to mine ore that is less rich in fuel. At prices up to $10 per pound, it is estimated that there are about 1,120,000 tons of uranium reserves. But according to the Atomic Industrial Forum, Inc., demand for uranium by atomic plants between now and the year 2000 will total almost 2.4 million tons, more than twice our reserves at $10 per pound. Present uranium prices and demand are not high enough to justify the huge exploration effort that long-range projections indicate will be required. The price of uranium will have to rise sharply, most likely to $30 per pound, before companies will make the necessary effort to extract 2.4 million tons. Supporters of nuclear power say that the cost of fuel has little effect on the cost of nuclear-generated electricity. Under the worst conditions a 100 percent increase in the cost of uranium would result in less than a 10 percent increase in the cost of electricity. We could get a much greater supply of nuclear fuel by allowing the price of uranium to rise to $50 per pound (which might result in a 50 percent increase in our electric bills). This would guarantee enough nuclear fuel to last us until well past the year 2025, by which time the breeder reactor program should be well established or other energy sources more available.

But atomic power need not only mean uranium. Other fossil materials are suitable as reactor fuel. Thorium is one. There is as much thorium around as uranium, and this fuel offers a definite advantage. It enables atomic power plants to operate at thermal efficiency 10 percent higher than uranium-fueled reactors because thorium is used in a high-temperature gas-cooled reactor (HTGR), a subject we will discuss shortly.

The final argument for atomic power is that present-day reactors (light-water reactors) are only an interim technology, so they need carry the electric energy load for only about 15 years. By 1990 a new technology called the breeder reactor will take over. Operating on still another type of nuclear fuel, plutonium, breeder reactors will be able to provide vast amounts of electricity for well over a thousand years. Fuel availability would not appear to be a serious problem for nuclear electric power.

SAFETY

The question of safety in nuclear power plants arouses strong emotions. Critics warn us that disastrous accidents can result from a mishap. It has been claimed by many critics, including members of the Sierra Club and the Union of Concerned Scientists, that if the supply of cooling water to the hot core of a nuclear reactor is disrupted accidentally, the uranium in the core would cause the water to rise to an extremely high temperature and melt the core. The hot molten atomic fuel would then melt through the pressure vessel and the outer containment structure into the ground (this is known as the China Syndrome), contaminating it with thousands of curies of radioactive material. This situation is referred to as "blowdown."

Because the question is so important, and because many prominent citizens and government officials were questioning the wisdom of building more atomic plants, the Atomic Energy Commission decided to make an in-depth study of the safety problem. It employed the services of Professor Norman Rasmussen of MIT and a team of 60 technicians. After years of research, analysis and theorizing, the Rasmussen group concluded that nuclear plants are eminently safe. The chances for a serious or catastrophic nuclear plant accident in any one year are one in 300 million—about the same as for being hit by a meteorite from space. According to the Rasmussen report, a person has a greater chance of being hit by lightning than being hurt by a nuclear power plant accident.

According to Dr. Dixy Lee Ray, head of the Atomic Energy Commission, atomic power plants have an extraordinary safety record. In 37 nuclear plants there were only 7 radiation-associated deaths in over 30 years. Furthermore, "no member of the public has ever been hurt in the operation of any commercial nuclear power plant from exposure to radioactive material."

The question of nuclear reactor safety was studied by the *Christian Science Monitor*, which sent questionnaires to 153 individuals across the country who had made technical presentations on reactor safety at American Nuclear Society meetings or had testified at Atomic Energy Commission hearings. The main question asked was, "How safe are nuclear reactors?"

Ninety-five percent of the respondents stated that present operating reactors are reasonably safe, barring operator error or sabotage. This view is upheld in strong terms by Herbert Kouts, director of the Atomic Energy Commission's Reactor Safety Division. According to Kouts, "Even with the 1000 reactors expected to be operating by the year 2000, it would be 1,000 to 10,000 years before any reactor might be expected to have an accident. . . . For the 100 to 200 years we expect to be using fissionable uranium before supplies run out we would expect never to have a catastrophic accident by an overwhelming probability factor."

Dr. R. Philip Hammond, a nuclear scientist with 30 years' experience, regards the danger of a nuclear reactor blowing up like an atom bomb as nil. It cannot happen, he insists, for a number of reasons. "A nuclear reactor has the wrong composition, the wrong surroundings and the wrong timing. The nuclear fuel contains substances which would prevent a bomb from igniting. Further, a bomb must be set off in clean surroundings, free from neutrons or it will preignite and shut itself off by thermal expansion. A nuclear reactor always has neutrons present and thus the wrong surroundings." Hammond goes on to explain that a bomb must be fired by pushing its parts together in a few millionths of a second. There is nothing in a reactor to give such speed. Reactors, Dr. Hammond insists, may experience many kinds of failures such as broken pumps and warped rods, which may be costly to repair, but they cannot become bombs.

Other scientists have joined in to defend nuclear power plant safety. They have argued that public health risks associated with the normal operation of electrical power plants (i.e., air pollution from fossil fuel plants and radioactivity from atomic plants) are modest at worst, and of these, nuclear power poses the smallest hazard; that the environmental impact of nuclear power is far less than that of coal-fired power and not much more than that of oil-fired generation; and that, considering fuel supplies and probable future prices, nuclear power is the most economical bulk electricity supplier.

Finally, Dr. Glenn T. Seaborg, former chairman of the Atomic Energy Commission, praises atomic power almost without reservations. "Nuclear power . . . represents the promise of an almost unlimited energy source. Commercial nuclear power plants have caused no injury to the public. There have been no radiation exposures to the public in excess of standards established. Nuclear plants are smokeless, quiet and esthetically attractive. They do not have unsightly stockpiles of fuel or noise and traffic associated with them. They do not put sulphur dioxide, oxides of nitrogen, carbon monoxide or even carbon dioxide into the atmosphere. . . . if our society continues to examine the relationship of nuclear power to the quality of our environment, these facts will become more widely recognized."

RADIATION

But what about radiation, that sinister killer that cannot be seen or heard? There is a tremendous amount of radiation present in nuclear plants. And we know that exposure to certain amounts of radiation can cause terminal cancer. But according to many nuclear scientists, there is nothing to worry about.

First of all, the way an atomic plant is built, no radiation can escape to the atmosphere. The massive structures surrounding the nuclear core stop almost all radiation that emanates from the nuclear fuel. Secondly, the Atomic Energy Commission had set stringent controls on radiation levels outside the plant, and also on the cooling water that comes out of the reactor and is emptied into local bodies of water. According to the commission, a person could drink the cooling water from an atomic plant all his life without exceeding the radiation limits set by the Federal Radiation Council. To verify this point, a study of the radioactive level of water around reactors on Lake Michigan was carried out by Environmental Research Groups, Inc. Their conclusion was that even if the worst leakage envisioned by the AEC were to occur, radioactivity in Lake Michigan's fish would not build up to a point higher than half the permissible level.

Perhaps the most dramatic rebuttal of critics citing the dangers of radiation was made by Bernard L. Cohen of the Scaife Nuclear Physics Laboratory at the University of Pittsburgh. Professor Cohen compares the risk of living next to an atomic plant with other more familiar risks. He claims that the added risk is about the same as driving 100 extra miles in one's lifetime, or swimming for one extra hour during one's lifetime, or smoking one cigarette every eight years, or being one hundredth of an ounce overweight, or living in a brick house rather than a wooden one for just six weeks, (bricks have more radiation than wood), or spending eight hours per year in Colorado or Wyoming, which have twice as much natural radiation as the other states. He goes on to say that in flying ten hours in an airplane exposes one to more radiation than living a lifetime near a nuclear power plant. Finally, he explains that a single dental X-ray exposes the body to more radiation than living near an atomic power plant for 20 years.

THEFT

Is there a danger that some terrorist organization or madman could steal atomic fuel, produce a bomb and blackmail the entire country? Opponents of atomic energy argue that we will have no peace of mind with a lot of atomic fuel, atomic waste (used fuel) and atomic plants around.

According to the utilities, power plant manufacturers and the Atomic Energy Commission, a lot of thought has been given to this problem. Once again, Dr. Philip Hammond reminds us that the uranium fuels used in present-day nuclear plants are of the wrong composition for bombs. It is also very unlikely that any organization would have the highly trained personnel and the huge amount of sophisticated equipment needed to produce weapons-grade uranium from nuclear fuel material. If any group had that capability (which would be extremely difficult to hide), it could work from uranium ore, which would be easier to acquire. According to Dr. Hammond, uranium fuels are no more dangerous than so much dynamite (a view held by other people in the field of nuclear power).

In two U.S. government reports, one from the General Accounting Office and one from the Atomic Energy Commission, the possibility that nuclear fuels might fall into the wrong hands was recognized. Their conclusion was that this danger could be averted by increased security at atomic facilities and during nuclear materials transport. The AEC report recommends that a counterintelligence network be set up to work with the FBI and CIA in developing intensive security methods for documenting and registering uranium and plutonium through each processing step. It would also organize a system of precise daily measurements to keep account of nuclear materials. Finally, its accounting procedures would be tight enough to detect losses of nuclear materials instantly.

Another point brought out by Professor David J. Rose of MIT is that security methods of handling nuclear fuel and waste have already been upgraded. Fuel is monitored during shipment in various ways. In addition, nuclear fuel is so difficult to handle that those who would steal it would be in the greatest danger. Regarding nuclear fuel security, Carl Walske, president of the Atomic Industrial Forum, states, "Is it possible to protect nuclear materials so that reasonable people will agree that any risks from diversion or sabotage are negligible? I believe it is and at a cost which, although high, need not be so high as to cripple the economics of nuclear power." Ralph W. Deuster, president of Nuclear Fuels Services and thus directly concerned with the cost of enforcing stringent regulations, maintains that the risk of nuclear materials theft is overdramatized, and that current measures are quite adequate. Finally, it might be observed that in over 30 years there has never been a successful theft of atomic material.

NUCLEAR WASTE

This subject has been brought up by many environmentalists and by prominent government leaders. Atomic power plants produce a great deal of long-lived radioactive waste which must be stored somewhere for many years. Eventually, the accumulated amounts will be huge. How can we feel safe knowing that this superdeadly poison may leak into our water supplies? Many nuclear scientists, leaders of the atomic power industry and representatives of the Atomic Energy Commission assure us that this is not the frightening situation it appears to be.

Frank K. Pittman, a Federal director of nuclear waste management and transportation, explains that there are very promising methods of storing nuclear waste. He states that highly radioactive fuel elements were experimentally stored in a deep salt mine formation for 18 months with no unusual or unexpected effects of radiation or heat on the salt, and were safely removed at the end of the experiment. According to the National Academy of Sciences' National Research Council, the best earthbound disposal sites may be impervious bedded salt formations 125 to 375 feet thick, located about 500 to 875 feet underground in geologically stable areas far from major population centers. Such a storage facility could safely contain radioactive waste for perhaps millions of years. Salt bed storage facilities are presently being studied at Oak Ridge National Laboratory. First, a shaft 1500 to 2000 feet deep is excavated. Then large rooms are carved out and holes drilled in the floor of the rooms. Special canisters containing nuclear waste are placed in these holes. The holes are then filled in. When the capacity of the storage area is reached, the entire room is filled. The heat and pressure from the radioactive waste recrystallizes the salt (causes it to form a solid layer around the canister); within not too many years the canisters containing the radioactive waste are effectively sealed in salt.

Adequate sites for this type of storage are not abundant, but right now the problem of waste storage does not seem to be pressing. According to Dr. Dixy Lee Ray, there is not enough nuclear waste from atomic power plants to be concerned about. Even by the year 2000 the anticipated amount of atomic waste (if it is all solidified) could easily be stacked on a tennis court. Dr. Ray says that nuclear waste storage "is the biggest non-problem we have, one really solvable through any one of a number of techniques." The AEC hopes to build a man-made facility to accommodate all the commercial waste generated for the rest of this century. Meanwhile, it is investigating "ultimate" solutions ranging from entombing wastes deep within the earth to rocketing them into outer space. According to Dr. Philip Hammond, the hardest thing for people to grasp is the extremely small volume of waste that will be produced; if all the electricity generated in the United States today came from atomic plants, and if these plants continued to produce at the present rate for 350 years, the total amount of nuclear waste (in solid form) could be stored in a room 200 feet long by 200 feet wide and 200 feet deep (or 8 million cubic feet). And the cost of treatment, transport and storage of nuclear waste is only one-thousandth of the cost for electricity.

Another method for disposing of nuclear waste is to place it in steel tanks embedded in concrete walls. Actually, this method has been safely used for almost 30 years. True, there have been some sizable leaks from tanks stored underground, but in each instance the tanks were located so as to prevent radioactive seepage from reaching the water table and finding its way into a community water system. One variation on this method is to drill holes in bedrock and place the canisters in these holes. Or, nuclear waste in liquid form could be piped down into lined underground cavities.

The cavity would be sealed when full and the heat generated by the waste would melt the cavity liner and surrounding rock, thereby sealing itself in permanently.

Some scientists believe that there are other ways to reduce the problem of radioactive waste. For example, nuclear waste can be divided into two categories: the fission products, which remain active for about 700 years (half-life of 30 years or less); and the actinides, the nuclear waste materials with active lives of hundreds of thousands of years (half-life of 25,000 years or more). It is technically possible to separate these two types of waste products. The actinides could be returned to nuclear reactors as fuel. The other waste materials would be far easier to handle, since their active life would be much shorter.

Another possibility is to turn atomic waste into other substances by a process called transmutation. Transmutation involves the bombardment of radioactive nuclides, causing them to become other elements with much shorter half-lives. These would present much less of a long-term storage problem. Other proposed solutions vary from dumping or burying the waste in the ocean to loading it on a huge Saturn rocket and shooting it deep into space or into the sun.

Much consideration has also been given to the use of radioactive waste as a heat source. Water would carry off the heat from the waste containers. Since radioactive waste generates temperatures of more than 1600° F., this hot water could be used for industrial purposes or even for commercial and residential heating. At the same time, the nuclear waste heat problem would be solved.

So there does appear to be a practical solution to every problem confronting the rapid advancement of nuclear power. But in order to fully appreciate the magnitude and complexity of atomic power, a brief view of the inner workings of nuclear reactors and nuclear power generating systems would be helpful.

THE HEART OF ATOMIC POWER

A good starting point is to look at what takes place in the basic building block of all atomic power plants—the center or nucleus of the atom. Although extremely small, the nucleus is the most massive part of the atom (see Figure 18), thousands of times bigger than the small electrons that revolve around it like miniature satellites. The nucleus is made up of two types of particles—positively charged protons and neutrons, which have no charge. Every atom except hydrogen has one or more protons and one or more neutrons. Uranium 235, the fuel used in nuclear

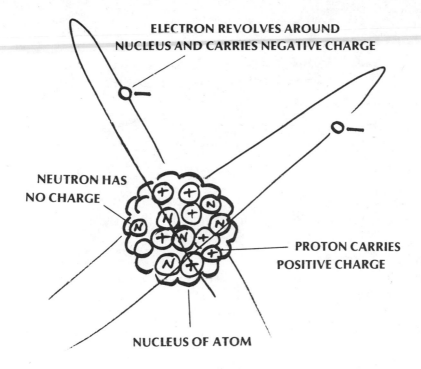

ELECTRON REVOLVES AROUND NUCLEUS AND CARRIES NEGATIVE CHARGE

NEUTRON HAS NO CHARGE

PROTON CARRIES POSITIVE CHARGE

NUCLEUS OF ATOM

FIGURE 18. The basic activity of nuclear power generation takes place in the nucleus of an atom.

plants, is made up of uranium atoms, which are not very stable. In other words, they have a tendency to suddenly break apart. This action is called spontaneous nuclear fission—or fission, for short. When fission occurs, the uranium usually splits into two equal pieces. It also produces a certain amount of heat (radiation) and it ejects some atomic particles, usually neutrons. This is the real key to atomic power (see Figure 19).

When a neutron is released, it flies off at high speed like a tiny bullet. If there are more uranium atoms around, it is likely to strike one. And since a uranium nucleus is unstable to start with, when it is hit by the neutron, it splits apart or fissions. The fissioned uranium nucleus also produces more heat and releases more neutrons, which proceed to split other atoms (see Figure 20). This series of actions is called a nuclear chain reaction. If there is enough uranium 235 and the chain reaction is not controlled, the result is an atomic explosion.

In a nuclear power plant uranium is located in a special assembly called the reactor core. As shown in Figure 21, the nuclear fuel is contained in steel rods, measuring about 1 inch in diameter and 12 feet in length, which are inserted into the reactor vessel. There are hundreds of

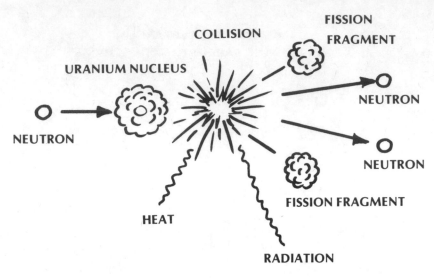

FIGURE 19. Nuclear fission generates great amounts of heat.

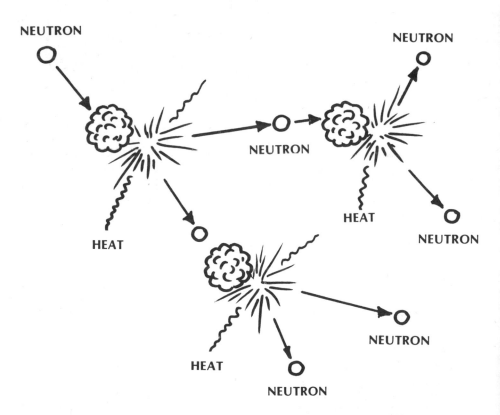

FIGURE 20. A neutron splitting one uranium atom results in more free neutrons, which proceed to split other atoms (nuclear chain reaction).

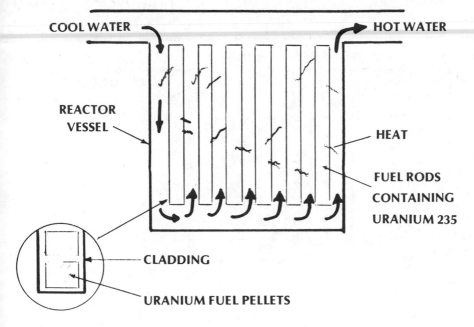

FIGURE 21. The heart of a nuclear power plant is the reactor core, which contains the nuclear fuel.

these rods in a reactor core. The uranium in the rods produces a great amount of heat, causing the rods to get very hot. If the heat were not carried off, the fuel rods would melt; however, huge amounts of water are introduced into the reactor core vessel to flow around the fuel rods and carry off the heat. This water is heated to about 600° F. as it leaves the vessel. The water also operates as a neutron moderator, which means that it helps the fission process to take place. When neutrons are released from uranium atoms, they travel at very high speeds; at these speeds, their chances of striking another uranium atom are not very good. But they must strike other uranium atoms if a chain reaction is to take place. So a moderator—in this case, water—is used to slow the neutrons down a little so they will have a better opportunity to collide with other uranium atoms. When slowed down, they are called thermal neutrons.

There is one potential problem with this arrangement: the reactor core might go out of control. Each time atoms are split or fissioned, other neutrons are released. Theoretically, this reaction could continue until such a huge amount of fission is taking place that the reactor core overheats and melts. But this does not happen because of the presence of reactor core control rods (see Figure 22). These rods contain special material, such as boron, which is able to absorb neutrons. Therefore when the control rods are totally inserted into the reactor core (between the fuel rods), they

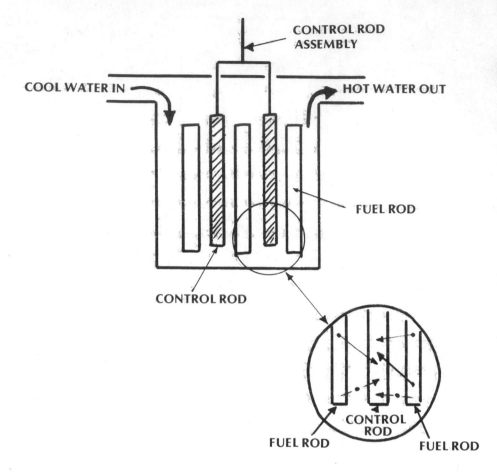

FIGURE 22. Control rods, containing neutron-absorbing material (such as boron), prevent fission in the reactor core from going out of control.

absorb many of the neutrons that are produced by the fissioning uranium. Since the neutrons that are absorbed by the control rods are unable to split other uranium atoms, the fission process within the reactor core is reduced to a low level. When the control rods are completely inserted, the reactor is virtually shut down. As the rods are extracted, there is less neutron-absorbing material between the fuel rods, and the chain reaction of the fission increases, resulting in a greater generation of heat in the core. The reactor core, then, is the heart of a nuclear electrical generating system since it producees all the heat required by the system. Although most atomic power plants are basically similar (in that the fission process produces heat), there are a number of different types of systems either presently in use or in the design stage.

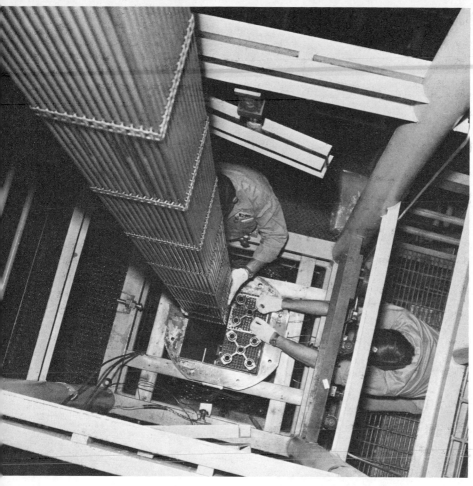

A nuclear fuel bundle (fuel rods) being pressure tested before actual installation in a power plant reactor core. (Courtesy Combustion Engineering Inc.)

LIGHT-WATER REACTOR

The most common type presently in operation is called a light-water reactor (LWR). It uses regular water (as opposed to heavy water or deuterium) as a coolant and as a moderator. Light-water reactors are of two types. The first is called a boiling-water reactor (BWR) (see Figure 23). In this type of power plant water is allowed to flow into the reactor core, where it is heated by the fuel rods. Since the water is not under high pressure (it is only at about 1000 PSI [pounds per square inch]), as soon as it heats up to its boiling point (between 500° and 600° F.), it is turned into steam. Actually, the water is allowed to boil right in the reactor core and the steam produced is sent to drive a turbine. The turbine, which operates somewhat like a toy pinwheel, is rotated by the steam, which is

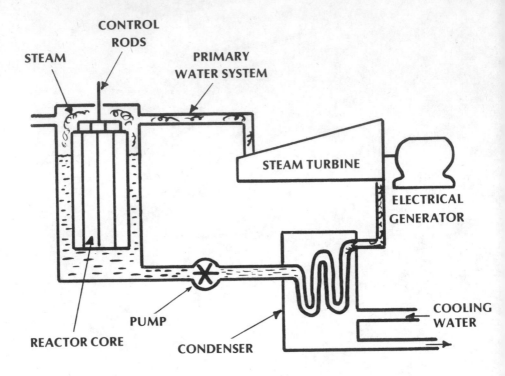

FIGURE 23. One of the simplest types of atomic power plants is the boiling-water reactor (BWR).

under a pressure of 1000 PSI. The rotating turbine is connected to an electrical generator, which actually produces the electricity.

Once the steam has passed through the turbine, its work is done, but it cannot be sent back to the reactor as steam. It must enter the reactor core as water so that it can collect a substantial amount of heat from the fuel rods. So the steam is passed through a tanklike assembly called a condenser. Cold water is passed through the condenser to absorb the heat from the steam, thereby condensing it to water. The primary system water (formerly steam) is then pumped back to the reactor core, where it is once again heated and turned into steam. All these steps form a continuous cycle as long as the reactor is in operation.

The second type of light-water reactor is called a pressurized-water reactor (PWR). In this type of power plant the water that is in the reactor core is kept under a pressure of 2500 PSI. As a result, even though the water reaches a temperature of 600° F., it does not boil; it is effectively superhot water and, as seen in Figure 24, it remains in a closed loop called the primary cooling system. In the PWR the hot water leaving the core is sent to a condenser-like device called a steam generator. Here the heat

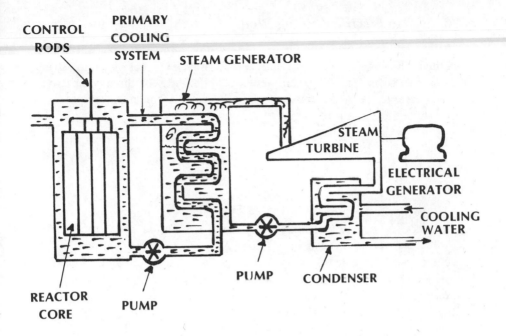

FIGURE 24. Another common type of nuclear power plant is the pressurized-water reactor (PWR).

from the primary cooling system is passed to a secondary water system which is not under high pressure. Subsequently, this water is turned into steam at 1000 PSI and sent to the steam turbine, after which the system operates in the same manner as the BWR. As Figure 24 shows, the water in the primary system is pumped back to the reactor core, where it is once again heated to 600° F. The steam leaving the turbine is cooled back to water by a condenser and returned to the steam generator.

OTHER TYPES OF REACTORS

Although most of the atomic power plants operating in the United States use light water as a coolant and a moderator, there are other types being used and being planned. One of these is the gas-cooled reactor. The main difference is that a gas such as carbon dioxide (under moderate pressure) is used as a coolant to extract heat from the reactor core. This heat is carried to a steam generator, where the hot gas is used to change water into steam.

There is a similar type of system called a high-temperature gas-cooled reactor (HTGR). In this case the cooling gas is usually helium and the system operates at a much higher temperature (about 1400° F.). With this method, the hot gas leaving the reactor core can either be used to heat water into steam (in the steam generator), or be sent directly to drive a steam turbine. A still different system, presently under development, is considered the reactor of the future.

WHERE 2 + 2 = 5

One of the problems with present-day atomic power plants (LWR) is that they operate on uranium 235 (235 U). Unfortunately, uranium ore contains only a small fraction (.7 percent) of uranium 235. Practically all ore (99.3 percent) consists of uranium 238, which is nonfissionable. So far we have not been able to use most of the uranium ore that we dig out of the ground. But scientists are working on a new type of nuclear reactor, an amazing machine that effectively produces a hundred times more atomic fuel than we presently have. This nuclear genie goes by the long name of liquid-metal fast-breeder reactor (LMFBR).

The operation of a LMFBR is quite different from that of a light-water reactor. First of all, in a LMFBR the fuel rods contain an element called plutonium 239 (239 Pu). This fuel is unstable (like uranium) and fissions. However, the neutrons that are created during the fissioning of an atom do not have to be slowed down (as in the case of uranium) by a moderator in order to create a chain reaction; the LMFBR works with fast neutrons. The nuclear core of this type of reactor has no moderating materials around its fuel rods. Figure 25 illustrates the inside of a LMFBR core. As in the case of other nuclear reactors, control rods are used to prevent the chain reaction from going out of control. But whereas reactors use either water or gas to cool the reactor core, the LMFBR uses liquid (molten) sodium to carry the heat away from the fuel rods. Sodium is used because it is capable of carrying a much greater amount of heat than either water or gas.

So far we have seen that the "liquid metal" in LMFBR refers to the coolant; the "fast" refers to the fact that fast neutrons are used in the reactor. Now we are left with the word "breeder," and this is where the important part of this system comes in. As seen in Figure 25, the plutonium fuel rods are surrounded by a blanket of nonfissionable uranium 238. What happens is that many of the neutrons (three out of every five) ejected from the fissioning 239 Pu fuel strike the uranium 238. When an atom of 238 U is struck by a fast neutron, it is converted into plutonium. Eventually, almost all of the uranium 238 in the blanket surrounding the fuel rods is converted into fissionable plutonium. Periodically, the blanket

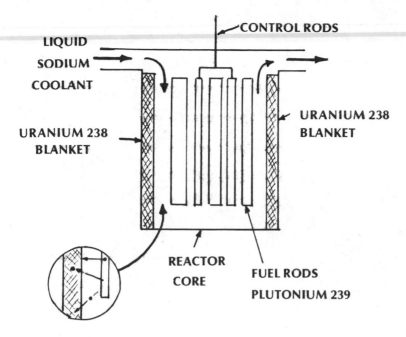

FIGURE 25. Liquid-metal fast-breeder reactors (LMFBR) will generate more fuel than they use.

material is removed and processed into fuel pellets to be used as fuel in a LMFBR. The time it takes the reactor to produce enough fuel to completely refuel itself and one other reactor is called the doubling time. If the plutonium produced by a reactor is simply stored, this doubling time will be about 16 years. However, it can be shortened if the plutonium generated by the reactor is quickly placed in another reactor (as fuel) so that it can begin, itself, to generate more plutonium from 238 U. This action could be compared to compound interest in a savings account (i.e., interest earning more interest).

One problem with the LMFBR is that the sodium in the primary cooling system after coming into contact with the fuel rods, becomes highly radioactive. To compensate for this condition, a secondary molten sodium cooling system is incorporated into the system (see Figure 26). While the secondary system carries off the heat from the primary system (through the heat exchanger), it never comes into contact with the highly radioactive fuel rods in the core. But the liquid sodium cannot be used to operate a steam turbine and so another cooling system is required, a primary water cooling system. The water enters the steam generator, where the hot sodium changes it to high-pressure steam. From this point on, the system operates similarly to a light-water reactor.

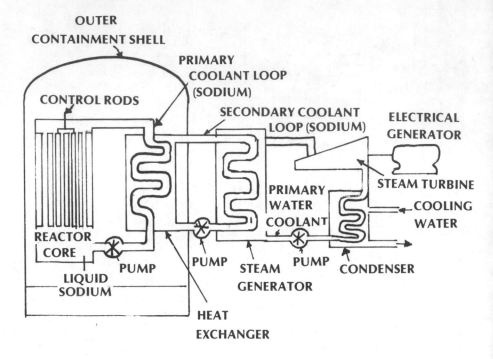

FIGURE 26. Liquid-metal fast-breeder reactor systems are more complex because they use hot liquid sodium as a coolant.

Besides producing additional fuel, the breeder reactor enjoys two other advantages over light-water reactors. First, because it operates at much higher temperatures (1200° F. vs. 600° F.), it uses heat more efficiently. The LWR thermal efficiency is only about 32 percent, whereas modern fossil-fuel power plants operate at a thermal efficiency of about 40 percent. The LMFBR also operates at 40 percent thermal efficiency, minimizing one of the criticisms of nuclear power plants.

The breeder reactor's second advantage is that it uses its fuel much more efficiently. A light-water reactor uses only 2 percent of the energy in its fuel; a LMFBR can use as much as 90 percent of the energy. So in this respect, the LMFBR is superior to the light-water reactor in meeting the power requirements of the future. And the way a number of large projects are proceeding, there will be a tremendous number of atomic power plants in our future.

Right now there are over 50 atomic power plants operating in the United States (all LWR except for two HTGRs). The former Atomic

Energy Commission forecasted that by the year 2000 about 1000 large (1000-MW) nuclear power plants will be in full operation, generating 60 percent of our electricity; many of these may be liquid-metal fast-breeder reactors. Even by 1980, it has been predicted, there will be about 132 nuclear plants supplying almost 20 percent of our electrical needs. This trend is being accelerated by two facts: our increasing demand for energy and our critical dependence on foreign oil. The OPEC nations have managed to keep their oil revenues at a lucrative level by lowering oil production and raising prices. Nor do they show any sign of disintegrating as a bloc; they will most likely remain unified until at least 1980. Meanwhile, other sources of fuel are not being developed at anywhere near the rate necessary to meet U.S. energy demands. For these reasons there is an intensive effort to construct more nuclear power plants and to complete the development of breeder reactors. This commitment is reflected in federal expenditures for energy research. In 1974 the research budget was $998 million. Of this, a whopping $530.5 million went for nuclear fission programs, almost all of it ($447.3 million) spent to develop the liquid-metal fast-breeder reactor. The 1975 nuclear power research and development budget was $1.5 billion, with the breeder reactor once again getting the lion's share.

It is hoped that these intensive programs will provide the technology to lift us out of the energy hole into which we have fallen. All the potential difficulties, according to nuclear advocates, are pretty well in hand. The thermal problems will be solved through more thermally efficient systems, through outflow systems and through the use of cooling towers and ponds. Location problems can be eliminated by placing nuclear plants three miles offshore, away from populated areas. Radioactivity is really not a problem because all nuclear facilities are closely monitored and include special mechanisms and extremely sensitive systems that can detect even microscopic traces of radioactive particles. The chances of a serious atomic plant accident are fantastically remote, the possibility of theft of weapons-grade nuclear fuel can be virtually eliminated with increased security and nuclear waste, at worst, poses very little threat to the environment. So, they claim, in a relatively short time the United States could provide for all its energy needs without suffering the environmental liabilities of conventional electric energy power sources.

But a closer look at nuclear power and its ramifications may reveal that this new energy source, to which we have made such a strong commitment, is not the ideal power source for our nation. In fact, it could well turn out that we are backing ourselves into a situation that is loaded with problems far more serious than the ones we presently face.

NOT ALL A BED OF ROSES

To start off with, nuclear power plants are extremely complex and challenge man's technological capabilities to their limits. In a conventional light-water reactor system there are between 400 and 500 variables to be measured and monitored—among them, fuel rod temperatures and position, radiation levels, seals, water temperatures and pressures, flow rates, nitrogen storage conditions, control and positions. All these systems must work to near perfection because there is little margin for error. An accident in an atomic plant can have disastrous and extremely long lasting effects. If radioactive materials were somehow to get into cooling water and be discarded to open waters, the effects of the discharge would continue for centuries. And with the introduction of the complex breeder reactors, the problems are intensified. In a breeder power plant there may be about 2000 conditions to be measured and monitored. Valves and relays, electronics components, cooling systems, chemical analysis, emergency core cooling systems, high-intensity radiation measurements, system isolation controls (between the molten sodium and the water)—how these and a thousand more variables perform will make the difference between a smooth-running system and a threatening one, or at best, an extremely costly shutdown of a giant power complex that could cost the public millions of dollars.

A dramatic example is the Palisades Nuclear Plant on Lake Michigan beach, six miles from South Haven. This 700-MW atomic power plant officially went into operation in December 1971. But because of technical problems, it operated at full power for only five months in a period of two and a half years. Then in January 1973 engineers discovered leaks of radioactive steam from one of the plant's two generators. The plant was shut down for seven weeks to plug the holes. In August 1973 more leaks were found, this time from the second generator. The plant remained shut down for over a year. An inspector of the system revealed that many of the tubes carrying water into the steam generator had corroded badly, causing a potentially dangerous situation. In May 1974 more steam leaks were discovered, leading to further delays in operation. These extensive delays were expensive—the cost of being shut down for one year was a tidy $36 million. And the Palisades plant is not unique. The Boston Edison Company's nuclear plant near Plymouth, Massachusetts, was out of action for over a year at the astronomical cost of $9 million per month. These extraordinary expenses ultimately translate into higher utility bills for local consumers. And we have only gotten started with nuclear power. The introduction of high (2000-MW) fast-breeder facilities will present us with far more complex problems.

All these problems are intensified when we consider placing these giant nuclear plants in an environment that is itself extremely complex— the ocean. The advocates of nuclear power technology have confidently decided that the surface of the ocean along the shorelines of the United States is the perfect place for atomic power stations. But even under the calmest conditions, the ocean is a dynamic realm, with its currents, tides, thermal gradients and large array of life forms. Under violent conditions, it can be a terrifying force. Although situating a huge atomic power plant in this environment has certain advantages (discussed previously), it also has a number of disadvantages—cost, for example. Today the price of a typical installation, including the breakwater, is $1 billion, but with inflation a floating nuclear power plant could easily cost $2 billion before it is finished, a very high price for one facility. The breakwater alone would cost around $250 million, and could easily reach $400 million. One could buy almost any piece of property in the country for that money.

Then there is the ever-present problem of heat pollution. Although representatives of the Offshore Power Systems Company have assured the Federal Power Commission that the heat discharged from a floating power plant will be quickly dissipated, no one can be certain what the effects of raising the local water temperature will be. About 3 billion gallons of water will be sucked in each day, passed through a heat exchanger (a condenser), where the water's temperature will be raised by 17° F., and then discharged to the open sea. Thus about five acres of ocean will become five degrees warmer. The result may be a growth of sea vegetation and the attraction of fish, which may disrupt the natural balance of aquatic life. More important, a major nuclear power plant accident could cause serious damage to the environment and threaten life.

Although the chances of such an accident are remote, the possibility must be considered. If there is a loss of sufficient cooling water, the nuclear core can overheat and melt through the containment vessel and even the prestressed concrete foundation. In a land-based plant the hot nuclear fuel would then flow into the ground, which is capable of containing it and absorbing a great deal of its radioactivity. But a floating nuclear power plant would pour hot uranium into the open sea—that is, tons of radioactive waste, containing enough strontium 90 to contaminate thousands of miles of water to levels above safe level. And we do not yet know how to decontaminate large bodies of water.

Added to these problems are the legal difficulties of a floating power plant. For example, could a state sell or lease an ocean area? What about property taxes? Would state law cover the installation? The answers to these and many other legal questions are still unknown.

Why should we subject ourselves to constant worry about the serious consequences of such a system when there are many other more acceptable alternatives? Why should we build 200 huge nuclear reactors off our shores, jam packed with tons of highly radioactive fuel, quite vulnerable not only to the risk occurrences of nature and accidents due to human error, but also to sabotage? Enemy destruction of floating nuclear plants would deprive us of an important source of energy, and at the same time seriously contaminate our coastlines with deadly radioactivity.

Apparently these stupendous problems have also occurred to the power utilities, because a number of their key executives now look upon floating atomic power plants with skepticism. By early 1975 the project seemed to be going under. Offshore Power Systems laid off about 60 percent of its work force and has postponed plans for constructing a large manufacturing plant near Jacksonville, Florida. Meanwhile, its only customer, New Jersey's Public Service Electric and Gas Company, has shelved its order for four reactor units.

Taking a good hard look at atomic power in general reveals imperfections that suggest it is far from the ideal solution to our long-range energy needs. Atomic fuel availability, for example, is not a foregone conclusion. In the early 1950s the general public was led to believe that as soon as atomic power plants started operating, we would have virtually unlimited energy. But it turned out that only a few elements could be used in atomic power plants. In fact, in present-day commercial atomic power plants only uranium 235 is being used as a fuel, and only a limited amount of it is readily accessible at reasonable prices. Because of the increasing demand for energy, we may begin to find ourselves hard pressed for fuel by the year 2020, less than half a century from now. Some nuclear power specialists maintain a major crash program of uranium exploration and production could meet the projected fuel requirements for the remainder of this century. But the availability is not the only difficulty with uranium ore.

Once coal is dug out of the ground, only a few, relatively simple processing steps (sizing, cleaning, etc.) are required to prepare it for use as a fuel. With gas and oil, the processes are slightly more complex. But with uranium, the preparation process is extremely complex and involves literally hundreds of operations. The uranium ore taken out of the ground contains less 1 percent of uranium 235, the necessary element. The ore has to go through a series of physical and chemical processes called milling to separate the uranium from rock and other materials in the ore. What is left over is a concentrated form of a uranium compound $U_3 O_8$ (uranium oxide). Unfortunately, milling releases some radiation to the atmosphere and generates small amounts of radioactive liquid and solid wastes.

The next step is to increase the concentration of uranium 235, which is required for the fission process to take place. This portion of the fuel preparation is called the enrichment process. For the present method of enrichment (called diffusion), the uranium must be in a gaseous form. So the uranium oxide is changed to another compound, uranium hexafluoride, (UF_6), which is easily gasified. This material is then shipped to a gaseous diffusion plant where, through a long series of operations, the uranium 235 is concentrated (enriched) from .7 percent to between 2 and 3 percent, which is the correct concentration for use in light-water reactors. A slight amount of radioactivity is released during the enrichment process. The enriched uranium is then shipped to a fuel fabrication facility, where it is converted to uranium dioxide (UO_2) and formed into small cylindrical fuel pellets. The pellets are then placed into metal fuel rods (cladding), which are the nuclear fuel rods used in the reactor core.

The complexity of this series of operations presents another potential limit to fuel availability for atomic power plants. As of January 1974 about 42 nuclear generating facilities were operating in the United States, another 56 were under construction; 101 were on order and 14 more were proposed in letters of intent. Since about 600 tons of fuel are required for initial loading (installing the first set of fuel rods) of each 1000 MW of electricity (the total generating capacity for these reactors is 204,000 MW), 122,400 tons of enriched uranium are required for initial loading. Furthermore, for each 1000 MW of capacity, an additional 200 tons of enriched uranium is needed for refueling over the life of a nuclear reactor (about 20 years). The problem is that we are running out of uranium enrichment capacity. It does no good to have the ore if we cannot produce usable fuel from it.

In May 1973 the Atomic Energy Commission reported that six new uranium enrichment plants, each with a capacity of 8.75 million tons per year (MTPY) of separative work units (SWU—a way of describing how much fuel is being produced), would be needed to meet future demands. None have been built yet. The cost of building these new plants, and of extensive exploration, mine development and milling, will probably surpass $20 billion by the late 1990s. Enrichment plants are especially expensive. A typical gaseous diffusion facility today would cost about $3 billion and require an entire 2000-MW electrical power plant for energy to run the enrichment equipment. Of course, there are other new and less costly methods, such as centrifuge techniques, jet membrane, and laser enrichment; but these approaches have yet to prove economical on a mass production scale. Another problem is that fuel processing and reprocessing (extracting fuel from old fuel rods) is now being turned over to private industry by the U.S. government, which, until now, had provided these services out of three huge facilities (Oak Ridge, Tennessee; Paducah, Kentucky; and Portsmouth, Ohio). But even the largest companies hesitate to take on such an expensive venture. A venture that requires an outlay of $3 billion and the training of thousands of new employees in the science (and art) of fuel enrichment is not very tempting. Many of these companies point to the disaster experienced by General Electric in its attempt to construct a nuclear spent-fuel recovery plant (extracting uranium from used fuel rods) near Chicago. GE spent six years and $64 million on the design and construction of this plant, which was to have been one of three new fuel facilities in operation in 1979. The plant does not work and will have to be virtually scrapped. To get the plant back on the track would require another four years and perhaps an additional $130 million. The mechanical and nuclear engineers were confident that they had designed the plant correctly, but when it was built, a lot of things went wrong. Plumbing clogged, parts broke down with a frightening regularity, equipment that was so radioactive it could never be touched by human hands turned out to be reparable only by human hands. General Electric has not ruled out the possibility of completely abandoning this useless plant. This disclosure shocked the utility industry, which was already beginning to worry about a national nuclear fuel shortage.

Another aspect of atomic power plants to be considered is thermal pollution. Nuclear plants are even worse offenders than fossil fuel plants in this area; because of their low thermal efficiency (32.5 percent), light-water nuclear reactors discharge about 50 percent more heat to the environment. There is a real possibility that we will have 2000 large nuclear power plants in operation around 2020, and their heat load will be taxing our environment to its limits. No one can predict the consequences of seriously intruding on the delicate balance of the thermal characteristics of nature. It is certainly not inconceivable that the reaction to such a thermal burden will be irreversible and totally destructive to society. Once again, it must be asked why we are insisting on taking such a needless gamble with an energy system that is inherently hostile to nature when there are other energy sources that would not add to the thermal load. The solutions to the heat problem, such as cooling ponds and wet and dry cooling towers, are very costly and, in the long run, would not reduce the heat load but simply distribute it. These methods could not defer for long the effects of an excessive heat load. In order to use the excess heat from atomic plants for commercial heating, the nuclear plant would have to be located relatively close to a populated area (the hot water would lose much of its heat energy if it had to travel a long distance through pipes), an unacceptable situation for reasons of safety. So excess heat, while not an overriding problem, is a serious negative aspect of nuclear power.

Radiation is a far more potentially dangerous situation. Radiation from nuclear fuels (uranium 235) can be divided into two categories: short-range and penetrating. Short-range radiation is in the form of alpha particles, which consist of helium nuclei (two neutrons and two protons) that are positively charged. These particles can easily do severe damage to internal body tissue. However, since they cannot penetrate human skin or a thin sheet of paper, or even a coat of paint, alpha particles have to be actually eaten or deposited inside the body to cause damage (radio poison). Penetrating radiation, on the other hand, is in the form of gamma rays, small quantities of matter which pass through most types of solid materials in the same manner as X-rays. To safely handle this type of radioactive material, one must work behind a shield equivalent to approximately 1000 feet of air, 20 feet of water, 10 feet of concrete, 3 feet of steel or 1 foot of a special uranium-type metal.

The decay or fission of uranium 235 also results in the formation of other materials called isotopes (these are transmutations of uranium), one of which is strontium 90. This fairly long-lived material can be absorbed in blades of grass. Cows grazing on the grass can accumulate strontium 90, which will be transferred to their milk. Anyone drinking this milk could get a large dose of this radioactive material. Strontium 90 settles in the bone and muscles, where radiation emanates for over 25 years, and can result in leukemia and other forms of cancer. Another product of uranium fission is the isotope cesium 137, which is also radioactive and settles in the bones and muscles. Another one, iodine 131, which has a half-life of only eight days, settles in the thyroid and salivary glands, damaging sensitive throat cells and triggering thyroid cancer. Finally, plutonium 239 (a mutation of uranium 238) is the most dangerous of the fission products and has a radioactive life of about a quarter of a million years. Exposure to plutonium radiation can result in body cell and chromosomal damage that may linger for generations.

The facts of radiation are pretty well accepted; what is in question is whether or not atomic power plants and their supporting facilities will add to the present level of natural or background radiation. Here we find a number of possibilities for accidental discharge of radioactive material to our environment.

Let us start with milling operations (we will consider the mined ore not to be a factor in radiation contamination). Once the ore has been fully separated, the proximity and concentration of the uranium begins to intensify the radiation of the raw material. Also, during this process it is possible to release trace amounts of radioactive material to the local environment, but these are probably negligible. The milled ore is then transferred to the enrichment facilities, which handle millions of tons of uranium each year and are not subject to the same strict regulations as are commercial nuclear power plants. Presently, there are three such processing plants; by the year 2000 there will be at least nine, increasing the probability of nuclear material being dumped into water systems, into the earth or (in gaseous form) into the atmosphere. Once it is fully processed, the enriched uranium (with fully active high-level radiation) must be shipped to a fuel production plant. During shipment the radioactive material is subject to a transportation accident, which could result in a discharge of this material to the environment. The fuel production plant and all its processes present another opportunity for an accident. The fuel rods produced in this plant must be shipped to the atomic power plant, where they are to be installed—another chance for a mishap. Finally, we get to the nuclear power plant itself where the fuel rods will be in operation for about a year. Before we discuss the power plant, we should take note of the fact that it is not the end of the line for the nuclear fuel. Once the fuel is spent (used up), it will be removed and transported to a fuel reprocessing center. Once again, there is an opportunity for an accident en route and a chance for leakage of radioactive waste at the reprocessing plant. In fact, it has been estimated that during the 1990s approximately 500 deliveries of highly radioactive nuclear fuel will occur. With that many shipments and with so many nuclear facilities in operation, the chances for radioactive leaks or spills will increase dramatically.

But let us return to the atomic power plant, for it is here that the greatest concentrations of atomic fuel will exist, under the most challenging conditions. The public has been assured by various private and government agencies that nuclear reactors are practically 100 percent (99.9999) safe. These agencies state that practically no radiation exists around reactors— that you can drink water coming out of reactor cooling systems—and insist that radiation levels would be safe even if all 1000 reactors that will be operating in the United States by the year 2000 were located in one place. However, in light of the available facts, these claims appear questionable.

There are two serious problems (aside from radiation) when dealing with uranium 235. One, it generates a tremendous amount of heat because of its radioactive state. Therefore the fuel rods must be made of heat-resistant steel, which can maintain its strength at temperatures of 1400° c. Two, hot uranium is extremely corrosive. It eats away most materials and will react with practically any metal. It literally penetrates the metal walls of fuel rods. So the rods must be made of special metals such as titanium, molybdenum, zirconium, tantalum and tungsten. The speed and penetration of uranium into the fuel rod walls depends on temperature. If the walls are coated with tungsten, they can withstand this penetration up to about 1100° c. However, if the fuel rod temperature rises above 1130° c., even tungsten fails to stop the hot radioactive uranium from penetrating the fuel rod walls.

Another challenge to the integrity of the fuel rods is pressure. The rods are normally tightly packed with uranium. In this condition they can withstand the pressure of the coolant (about 2500 PSI for PWR) surrounding the reactor core. But under certain conditions the fuel within the rods contracts and leaves spaces between the fuel pellets. This can lead to the collapse of the fuel rod walls and cracks in the wall structure. Once this happens, highly radioactive uranium and plutonium (which is formed in reactor rods) can leak into the cooling water. This, in turn, carries radioactive waste into the steam generator and to the condenser, which receives primary cooling water from local bodies of water (lakes, rivers or oceans). What are the chances of such a mishap?

According to nuclear experts, they are very low. According to fact, they are excellent. The Robert E. Ginna reactor in Rochester, New York, experienced such a mishap. At nuclear reactors in Hartsville, South Carolina, and Two Creeks, Wisconsin, hot spots on fuel rods (caused by intense radiation) were discovered. Structural failure also led to radioactive steam leaks at the Palisades nuclear plant in Michigan. Furthermore, official government reports show a number of releases of radioactivity above the permissible limits (but still within safety limits). At the Quad Cities' nuclear plant near Cordova, Illinois, 11 releases of gaseous iodine 131 were traced to radioactive steam leaks. Many other cases of structural failure, accidents, poor quality control and plain human error can be cited.

As the number of nuclear plants increases (from the present 56 or so to possibly 2000), and as their size increases (from about 700 MW to 1000 or 2000 MW), the chances of an accident due to structural failure or an operation mishap will increase correspondingly, to say nothing of the natural deterioration of any mechanical system. The leakage of radioactive material to our ecological system is a truly frightening prospect. Scientists are beginning to learn a great deal about the effects of radiation on human beings, especially children. Ernest J. Ternglass, professor of radiation physics at the University of Pittsburgh, states, "The unanticipated genetic effect of strontium 90 has become evident from an increase in the incidence of infant mortality along the path of the fallout cloud from the first atomic test in New Mexico in 1945, and from detailed correlation of state-by-state infant mortality excesses with yearly changes of strontium 90 levels in milk." He goes on to say that infants are most susceptible to the effects of radiation, which often means leukemia (this discovery was made by Dr. Alice Stewart of Oxford University). To this, Barry Commoner adds, "Despite earlier uncertainties, it is now generally accepted by the scientific community that the harmful effects of radiation are proportional to the radiation dose." In other words, even a small amount of radiation in addition to the amount we receive naturally can cause biological harm.

These facts should be kept in mind while reflecting on the results of a study done on the Columbia River in Oregon. Tests showed that though the radioactive level of the water was rather low, the effects on the natural food chain were considerable. The plankton in the river were 2000 times more radioactive than the water. The fish that fed on the plankton were about 15,000 times more radioactive than the water. The waterfowl that ate insects from the river were 500,000 times more radioactive. Finally, the eggs of the birds that lived on the Columbia River were one million times more radioactive than the water. Where is all this radiation coming from? The source is leakage from the Hanford Atomic Products facilities at Richland, Washington.

But the Hanford facility does not have a monopoly on leakage; around May 8, 1974, small quantities of plutonium 238 were discovered in the sediment in a canal near the Mound Laboratory in Miamisburg, Ohio. This isotope is one of the most poisonous known. It can cause almost instant death if breathed into the lungs. Even as nuclear scientists are claiming that radioactive leaks will not occur because of sophisticated controls and extremely sensitive detection systems, engineers at Mound Laboratory are hard at work trying to find out how the radioactive plutonium escaped from the process steam and leaked undetected. Meanwhile, at the government's Savannah River processing plant failure in a processing line resulted in the release of radioactive tritium gas into the atmosphere. Federal officials have been investigating the area around the plant to determine if water, milk or vegetation has been contaminated.

It is a common mistake to consider the effects of radioactive pollution as localized. The oceanographic scientists at Woods Hole Institute will tell you differently. According to Arthur Maxwell, provost of the Institute, the things that chemists find in a glass of seawater collected in remote corners of ocean include traces of pesticides and a range of man-made radioactive materials. Edward Goldberg, of the Scripps Institution of Oceanography in La Jolla, California, states, "We may be playing a game of scientific roulette. Nobody knows when these measurable traces will become dangerous. It could be hundreds of years. It could be far less. What we do know is that once they are there, they stay there."

The underlying problem is that we do not know enough about the subtle effects of small doses of radiation. Is it true that small amounts of radioactivity may safely be added to our environment? Or should we add none at all? Should we not be more concerned about the natural processes that tend to concentrate radioactive material in man's food chain? Certainly, if we can possibly build other systems of generating energy that don't add to our present radioactive burden, we should do so.

It has been pointed out that present levels of nuclear radiation from power plants are much lower than background or natural levels. But man has been living with the natural levels for millions of years and the human body has necessarily adjusted to them. It cannot be assumed that even small increases in radioactivity (which is cumulative in the body) will not have adverse effects. So once again we must ask, why risk a situation that might destroy our society? There are a number of alternatives to atomic power, which, if approached with equal vigor, could relieve the present energy crisis without substituting a health crisis.

We have been told by nuclear energy proponents that the chances of a major nuclear accident involving a steam explosion are extremely slim: one in 300 million. A number of independent scientists question the veracity of that statement and the mathematical procedures that led to it. If nuclear power plants are so safe, why do the utilities insist that the United States government retain the Price-Anderson Act of 1957, which exempts electrical power companies engaged in the business of nuclear power from public liability? If there is absolutely no chance of a nuclear explosion, if there is no chance of a primary cooling system failing and giving the surrounding inhabitants a lethal dose of radiation, why won't the electric utility companies carry their own liability policies? The answer is that American insurance firms (and even Lloyd's of London) would not touch these companies with a ten-foot pole.

Philip L. Rittenhouse of the AEC's Oak Ridge National Laboratory has testified that many nuclear safety experts have privately stated their concerns about safety standards and equipment. Their concerns are understandable when one thinks about the Humboldt Nuclear Power Plant built directly over a geological fault in California, one of the most seismically active areas in the world. One good earthquake and—*wham!* Even federal officials are really worried about that one.

Many other eminent scientists—for example, Dr. J. W. Gofman of Berkeley and Dr. A. R. Tamplin of Livermore (the authors of *Poisoned Power*); Dr. Henry W. Kendall, professor of physics at MIT; Dr. Ian A. Forbes, a nuclear engineer at the Lowell Institute of Technology; Dr. Gerold M. Lowenstein of the University of California's Radioactivity Research Center, who holds degrees as a medical doctor and a nuclear physicist; and even H-bomb pioneer Edward Teller—have very grave reservations about rushing into a large nuclear power plant program.

What would happen if some unforeseen incident (a jumbo jet crash, a serious earthquake or a terrorist's bomb) were to cause the primary cooling system in a plant to become inoperative? According to Dr. Henry Kendall, the nuclear core would become a molten mass so hot it would melt through anything guarding it. The contact of random water flow with the superhot fuel would result in a rapid buildup of radioactive steam or an actual steam explosion that could rupture the outer nuclear reactor containment wall. Thus a cloud of steam would be discharged to the atmosphere, carrying a deadly cargo of high-level radioactivity. Many people exposed to this cloud would be dead within two weeks. Could this really happen? Practically impossible, insist the defenders of atomic power. But once again, the facts make the neutral observer skeptical.

According to an article in *U.S. News and World Report* (June 10, 1974), a spokesman for the Atomic Energy Commission said that in 1973 utilities in the United States reported 861 "abnormal occurrences" in their nuclear facilities and 43 percent of these incidents had "potential safety significance." The article cites an example that approached catastrophe. In 1973 the Vermont Yankee nuclear plant near Vernon was ordered out of commission because of the presence of an internal vibration. While workers were locating the trouble, two control rods (which prevent the nuclear fuel from becoming overactive) were accidentally pulled at the same time, setting off a chain reaction in the core. While this potentially catastrophic accident was in progress, the heavy steel protective cover was off the reactor core.

Dr. Kendall has publically stated that "we were in possession of a Safety Review prepared by the Regulatory Division of the A.E.C. which said that reactor program was besieged with various safety problems." According to the report, 200 nuclear plant malfunctions were primarily due to poor design and/or error, improper maintenance, administration deficiencies and random failure.

One bone of contention is the lack of independent confirmation of safety research being conducted by nuclear reactor manufacturers. When questioned about atomic power plant emergency cooling system dependability, 80 percent of the national laboratories and 56 percent of the respondents from academic institutions felt that sufficient dependability has not yet been demonstrated. With the fast-breeder reactor, the question of safety becomes even more critical.

Great concern regarding our rapid plunge into breeder reactor technology has been voiced by a number of eminent scientists. Their fears were summarized in a statement put out by Nobel Laureates Linus Pauling, Harold Urey, James Watson and George Wald: "The [LMFBR] reactors cooling system will utilize liquid sodium, which is highly reactive and burns on contact with air or water. Breeder reactors are inherently more difficult to control than today's commercial fission reactors, they operate closer to the melting point of their structural material, and they generate and use much larger quantities of plutonium. Plutonium has a half-life of 24,000 years and is one of the most toxic substances known to man." There will be about 2500 pounds of plutonium in a breeder reactor, enough to make dozens of Hiroshima-sized bombs. The radiation inside a breeder reactor is so intense that it can cause structural damage and make steel swell. A serious accident involving a LMFBR has already occurred. One of the pilot breeder reactors operated by the AEC suffered a partial melting of its fuel while an experiment was being conducted in 1955. The reactor was destroyed. A similar but less serious accident occurred in 1966. And these accidents took place under the watchful eyes of trained scientists. The chances of such an accident when 1000 or even 2000 nuclear reactors are in constant operation, and attended by less qualified personnel, are frightfully greater.

So the great odds against such an accident, determined by the Rasmussen report as one in 300 million, are not very comforting. The chances of drawing a full house in poker may be one in a thousand, but that does not mean one will have to wait a thousand hands for it to happen. One might very well draw the full house in the third hand of cards. Similarly, we cannot assume the mathematical odds will protect us from a serious accident for ten thousand years; we may see one in ten years (after which another one may not occur for ten thousand years). Why should we bequest to future generations the awful responsibility of standing watch over a nationwide network of facilities that require unwavering vigilance and flawless security merely to generate a form of energy that can be produced by a half dozen methods that work in perfect harmony with nature?

But the leakage of radioactive gases and liquids from atomic plants and processing facilities pales in comparison with an even more potentially dangerous situation—long-term storage of large amounts of highly radioactive waste. The statement of AEC Chairman Dixy Lee Ray, that nuclear waste "is the biggest nonproblem we have," one that is "readily solvable" through any number of techniques, reflects the thinking of many scientists and leaders in the nuclear power business. According to the Council on Environmental Quality, all the activities of nuclear power are closely monitored by supersensitive electronic instruments. We have no problem, says Ray. Well, if the existence of 85 million gallons of highly radioactive waste from the military and 600,000 gallons of highly radioactive waste from atomic plants is no problem; if trying to store 85.6 million gallons of extremely dangerous radioactive waste that can eat its way through steel and concrete is no problem; if the necessity of storing this supercorrosive, extremely toxic, heat-generating radioactive material for a quarter of a million years is no problem—then we can all enjoy the coming age of nuclear power.

Radioactive waste contains all the most dangerous short- and long-lived atomic isotopes such as strontium 90, cesium 137, tritium, and—most frightening of all—plutonium, one of the most toxic materials known to man. It is true that presently we do not have a tremendous amount of high-level radioactive commercial waste (600,000 gallons). But it is also true that we now have only 40 nuclear reactors, and many of them are in the 600- to 800-MW range. By the year 2000 we will have about 1000 large reactors (1000 to 2000 MW) in full operation, which will multiply the waste problem considerably. Looking ahead to 2020, as many as 2000 fast-breeder reactors, with capacities of 2000 to 3000 MW, may be running at full blast. About 10 billion pounds of uranium will be required to fuel the nuclear power plants built by 2000, and another 20 billion pounds will be needed for the reactors built by 2020. (And these figures are only for our own nation.) Therefore by the year 2000 we will have to safely store about 370 million pounds of long-lived, highly radioactive atomic waste. Add to this 120 billion pounds of waste by the year 2020. And after that, keep adding 60,000 pounds for each 1000 MW of capacity (by then we may have about 3 million MW of capacity—generating about 180 million pounds of nuclear waste each year). Maybe Dixy Lee Ray is not concerned about what to do with this vast amount of killer material, but other highly qualified citizens are. We just do not know any safe way of containing nuclear waste for the eons that are required. We cannot even be sure that our present form of government or society will last that long—to say nothing of the natural elements.

The words of Dr. Ray continue to ring out: "We do not have a problem with nuclear waste." What makes them sound so hollow are two serious situations. First, we don't know nearly enough about the nature of radioactive materials, especially plutonium. The nuclear experts are just beginning to learn about the complex and subtle effects of a plutonium leak. Although it has long been known that large spills would destroy surrounding vegetation, scientists had in the past considered small plutonium leaks relatively harmless. Recent laboratory research has revealed, however, that plants are capable of absorbing plutonium from the soil. According to Dr. Raymond R. Wilding and Thomas R. Garland of Pacific Northwest Laboratories, researchers who "proved" that plutonium was inert in the soil may have been misled by inferior measuring techniques. Drs. Wilding and Garland, using a technique that enabled them to measure extremely low concentrations of plutonium, discovered quantities of plutonium that had been absorbed into the roots of plants. Studies showed furthermore, that in addition to absorbing plutonium into their roots, plants distribute small amounts of plutonium further down into the soil through their root systems. Continued study of this process is extremely important. If a plutonium spill were to occur, root plants in the area (such as potatoes, beets, radishes, carrots) might absorb doses of plutonium that may be hazardous to human beings. Not only must this process be completely understood, but the effects of low levels of plutonium on the human body have yet to be fully evaluated. We as a nation are rushing headlong into nuclear power—and more specifically, plutonium fuel reactors—while we are still operating on health standards for plutonium established in 1949, when we knew much less about this killer isotope. In 1974 the Natural Resources Defense Council, a highly respected environmental organization, issued a technical paper that stated that current exposure limits of airborne plutonium are much too high—in fact, 100,000 times too high. The eminent health physicist, Karl Z. Morgan of the Georgia Institute of Technology, believes that current allowable plutonium levels should be reduced by a factor of 40 or 50. The more one looks into the field of atomic power, the more one becomes convinced that we are running along a precipice, blindfolded. Yet Dr. Dixy Lee Ray insists "we do not have a nuclear waste problem."

The second reason Dr. Ray's words sound hollow is that we still have no solution to the problem of long-term radioactive waste storage. It has been recommended that we store this waste beneath the ground in salt mines or in bedrock. But we are not talking about ten years, or a hundred years or even a thousand years. We are talking about safely containing this corrosive material for a quarter of a million years—50 times longer than the recorded history of man. Who can predict what seismic or meteorologic changes will take place in the distant future? With 1.8 billion pounds of highly toxic waste being added every ten years in our own country—and this figure does not include the waste from reactors that would be built after 2020—together with billions of pounds from other nations, the world will constantly be on the brink of disaster. What will we do with this waste for the next 249,975 years?

One suggestion is that society create a special "priesthood" dedicated to a ceaseless vigilance over nuclear waste throughout the centuries. Another is that we construct huge storage vaults that would dwarf the great pyramids, last for millennia and always remain conspicuous so that no one would ever forget their purpose.

But the ERDA (the Energy Research and Development Administration, which replaced the Atomic Energy Commission in 1974) does not concur with either of these suggestions. Instead, it believes atomic waste materials should be stored in large steel and concrete tanks, under close observation, so no highly toxic materials will be discharged into our environment. Meanwhile, the stuff continues to pile up and the suggestions keep coming in. One idea is to dump the nuclear waste into the sea. This proposal should be rejected, according to Charles Hollister, a scientist at the Woods Hole Oceanographic Institution, because even if the radioactive wastes are diluted by all the water of all the world's oceans, "at projected production rates, all the oceans would be polluted by the year 2000." Dr.

Hollister goes on to explain that the ocean floor is not a good place to dump nuclear waste. Dumping in the seismically active "ring of fire" around the Pacific, or any other active area, would be foolhardy. Dumping in a potential resource area would be just as bad—imagine an underwater manganese miner plowing through a keg filled with hot plutonium. Neither would it be wise to try to dump waste-filled containers near plate boundary areas (huge continent-sized sections of the earth's crust that are in continuous motion). We do not know enough about ocean floor movements or about the rates, directions or other characteristics of ocean currents to make the sea a feasible dumping ground.

Similar problems negate the idea of burying containers of nuclear waste in the polar ice caps and allowing them to melt to the bottom. We do not know the long-range results of such a course of action, and we could not be certain of the containers' eventual destination. Also, it would be nearly impossible to get to the containers for purposes of inspection, and if a serious leak were to occur, we would be virtually powerless to stop it—if we were able to detect it. Furthermore, massive storage of nuclear waste at the poles would require international agreements, which would be difficult to obtain.

Since the earth does not seem to have any long-term nuclear waste storage facilities, why not pack the stuff on a rocket and shoot it into space? Despite the appeal of this idea, it is impractical for two reasons. First, it would be extremely expensive. It costs about $1000 to put one pound into earth-escape orbit. And with all the nuclear waste we will be accumulating, the bill would be staggering. But even if we could afford the cost (which would ultimately be paid by the consumer), there is the matter of safety. It is quite difficult to attain a very precise orbit. We could never be sure that a payload of deadly plutonium would not eventually curve around in space, head back to earth and burn up in the atmosphere, spreading its deadly cargo over our heads. Of course, that is assuming in the first place that the launch would be successful. Consider a huge Saturn, or a Titan, on the launch pad with its load of plutonium. A thousand little things could go wrong soon after liftoff, requiring that the launch vehicle be destroyed in flight. The concept of storing nuclear waste in space is appealing only in theory. The question remains—what can we do with all our nuclear waste?

According to some nuclear scientists, we can simply keep collecting it and wait for an eventual solution. James Schlesinger, when chairman of the AEC, described the problem quite clearly: "If you have the feeling that the AEC is threshing about, you are correct." But then he went on to explain that the worst that can happen is that no simple, inexpensive way will be found to remove the waste from our environment. In that case we would have to keep it within our environment, under constant surveillance. "If there is surveillance," he explains, "then there is no risk."

No risk! Confident words that parrot Dr. Ray's "no problem." Unfortunately, this confidence is unwarranted. Leaks of radioactive steam have already been cited; they alone cause skepticism in about the contain ability of radioactivity. What makes one even more skeptical is the lack of major research in the area of atomic waste. Senator Frank Church of Idaho states that while the AEC has spent many billions on the production side of the nuclear industry over the past 25 years, in all that time only $50 million has been spent on the problem of waste storage. In fiscal 1975 the agency spent only about 8 percent of its $385 million budget for civilian reactor research. Obviously, the nuclear industry is far more interested in getting atomic power plants into operation than in solving the problem of nuclear waste. Finally, here are a few incidents that should turn the most optimistic person into a skeptic.

Around April 20, 1973, at the giant Hanford nuclear reservation near Richland, Washington, Tank 106T, one of the many 500,000-gallon capacity tanks of highly radioactive liquid waste, began to leak. The deadly liquid gurgled out, undetected, for an entire day. How is it that all those sophisticated electronic detecting instruments, supposedly capable of pinpointing a microscopic leak of radioactive material, did not sound the alarm? The liquid continued to escape, gallon after gallon, for five days. During this time more liquid was being pumped into 106T. Only when pumping operations were stopped did it become obvious that something was wrong because the level of waste in the tank was dropping—an extraordinary way to find out that one of the deadliest substances on earth is pouring into the ground. Surely, one would think, emergency procedures started instantly on April 25 or April 26 or April 30. Unbelievable as it may sound, it was not until June 8, seven weeks later, that officials of the Atlantic Richfield Hanford Company (which is responsible for nuclear waste management) realized what had happened and ordered emergency pumping out of Tank 106T. By that time 115,000 gallons of highly toxic nuclear waste had escaped from the tank and poured into the ground, including 14,000 curies of deadly strontium 90 and 40,000 curies of cesium 137. One would imagine that after such a serious accident, officials would have taken great pains to ensure that it would not recur. But, astonishingly enough, it did. Early in 1974, at the same installation, about 2500 gallons of nuclear waste leaked out of a tank. Then on May 3 and 4 still more leaked out—this time approximately 500 to 2000 gallons.

A hair-raising situation exists at the Hanford facility. "Intermediate-level" waste is simply pumped into a bottomless concrete "crib" and allowed to seep into the soil. There are those who fear that these radioactive materials will eventually find their way to the Columbia River, which actually flows through the AEC facility. Even the AEC is concerned. Beneath one of the Hanford intermediate-level waste cribs so much plutonium has accumulated that the agency experts fear the pile may go critical (i.e., initiate an uncontrolled chain reaction, as in a nuclear bomb). Steps are being taken to remove much of this material. To date 17 leaks have been experienced at Hanford and 7 at the Savannah River facility. And Dixy Lee Ray says we have no problem.

Perhaps our biggest problem is the attitude of many scientists and nuclear engineers. In the editorial section of *Electronic Design* (October 11, 1973 issue) it was reported that "the scientists and engineers in nuclear plants who were interviewed feel that nuclear waste is more a public-acceptance problem today than one of disposal. 'We have until 1982 to decide on a solution,' an industry spokesman notes. That year is the one set by the Atomic Energy Commission for providing a permanent national disposal system for nonreclaimable waste from commercial reactors." What is going to happen in 1982? Will a magical solution suddenly manifest itself? Not likely. And meanwhile, atomic waste will pile up and we will be further along in the construction of bigger atomic power plants.

Heat pollution, radiation leakage, and nuclear waste do not exhaust the subject of atomic power problems. There remains one more danger, a danger that may in the long run prove the most devastating—nuclear terrorism.

In 1961 there were 200,000 shipments of radioactive material. Most of the nonmilitary shipments were for medical and other scientific uses. By 1971, when only 17 atomic power plants were in operation, the number of radioactive material shipments had quadrupled to 800,000. Shipment patterns are complex. For example, radioisotopes are produced in nuclear research and production reactors in Tennessee, Virginia and Georgia; uranium ore is mined in Utah, Colorado and New Mexico; materials are processed by radiochemical facilities in New Jersey, Texas, and California; spent nuclear fuel from reactors is processed in Illinois, New York and South Carolina; and nuclear waste is buried in half a dozen states. This was the picture when there were only 17 atomic plants. Now there are over three times that number and the picture is becoming more complex. By the year 2020 the matrix of power plants (2000 of them), milling plants, enrichment facilities, reprocessing plants and nuclear waste storage locations will result in a fantastically complex pattern of transportation. These highly radioactive nuclear materials must be protected from falling into the wrong hands. Security will be one of the toughest problems faced by the nuclear industry.

The idea that large amounts of extremely dangerous materials are within the reach of small numbers of determined terrorists is quite sobering, especially since a mere 18 pounds of plutonium is enough to build a small atomic bomb. American nuclear power capacity is expected to triple by 1980 and foreign capacity to increase by 800 percent, and the supply of plutonium will rise dramatically. So the material will be susceptible to theft, and even the instructions required to build a bomb are within easy reach. In an AEC report (the Rosenbaum report) it is pointed out that "precise and accurate instructions for the manufacture of such a bomb are available in unclassified literature. . . . The potential harm to the public from the explosion of an illicitly made nuclear weapon is greater than that from any plausible power plant accident." Dr. Daniel Kleitman, professor of mathematics at MIT, and one of the AEC report investigators, explains that "Future quantities of the stuff [plutonium] will be so large that people involved in processing [i.e., spent-fuel reprocessing] will (eventually) take the loose attitude that one percent loss is not significant. But when that one percent grows to 10 or 20 kilograms, enough to build a small bomb—it is significant indeed."

The soaring prices of fossil fuels will force most nations to turn to atomic power if alternate technologies are not developed. The United States is supplying technology and nuclear reactors to Egypt. These reactors will produce plutonium as a by-product—plutonium that can be used for bombs. And so it goes. Everywhere one looks, there will be nuclear reactors—reactors by the hundreds, then by the thousands—and each one of them will be producing plutonium. Arms experts Drs. Mason Willrich and Theodore B. Taylor point out that one terrorist group with one nuclear bomb could blackmail an entire metropolis. (A nuclear bomb weighing less than 22 pounds could destroy a medium-sized city.) Dr. Willrich fears that eventually a black market in fissionable materials could develop, with organized crime stealing from private nuclear facilities and selling to individuals or governments.

It has been suggested by some people, including Mr. Paul Turner of the Atomic Industrial Forum, that poorer nations may not choose to build "A" bombs. "They probably would feel they do not want to, or are not capable of making the enormous economic and technical commitment that is required to develop this weapons capability because it would draw resources from other important avenues they might choose to follow." But when one thinks of India with its starving millions, Mr. Turner's words become meaningless. Now that India has the atom bomb it would not be surprising if one or more of its neighbors (such as Pakistan) felt compelled to balance the scale of power. With nation after nation wanting to balance the power scale, there will be a frightening amount of destructive power in the hands of many political leaders—not all of whom can be counted on to show restraint under trying conditions. Even worse are groups or individuals who are fully convinced that they are justified in changing national political structures at any price. For these people there are no failsafe controls.

Even if an individual or group could not fabricate a nuclear bomb, the mere possession of plutonium could still cause a great amount of damage. If the plutonium were encased in a large bomb (built with nitroglycerin or other high explosives), the explosion (and wind) would scatter it so that it could contaminate huge areas for centuries. Keep in mind that inhaling a small speck of plutonium can cause cancer, and the scattering of one year's waste of an atom plant would contaminate an area of 10,000 square miles. By the year 2000 there will be enough plutonium stored in the United States to kill 100 trillion people.

Gone are the optimistic dreams and fantasies of the early 1950s when we were told that atomic power meant practically limitless clean energy and the solution of many of our social and economic ills without our having to pay a costly price in return. As it turns out, the price will be extremely high in many ways: in heat pollution; in the ever-constant fear of nuclear power plant accidents; in increased radiation; in the terrible burden of being caretaker to huge amounts of deadly radioactive waste which must be guarded for millennia; and the constant risk of mass destruction from nuclear terrorists.

What will we do two or three decades from now, when our economic system is based on nuclear power, if it turns out to be a nightmare? Future generations will be stuck with our deadly blunder. We are back to the old question: Why are we insisting on taking this course of action to the near exclusion of other courses that could safely satisfy our energy needs? Supporters of nuclear power tell us that other energy sources cannot meet future demands. That is a moot point at best. If all the billions of dollars and engineering and research personnel that have been put into the nuclear power program—in particular, the breeder reactor program—had been applied to other sources such as solar, geothermal, wind, tidal and fusion power, the nation would be well on the way to solving its energy crisis without having to pay such a terrible price.

In 1974 the Ford Foundation concluded a four-year study on energy in the United States and made the following recommendations:

• The federal government's open-ended commitment to the liquid-metal fast-breeder reactor should be retracted. The government should put up no more than 25 percent of such demonstration projects. The National Academy of Sciences should study the feasibility of proceeding *at all* with the LMFBR.

• No new nuclear power plants should be purchased, but those under construction should be finished.

The study assumes that nuclear power is risky and points out that if power demand growth were held to 2 percent for the next decade, followed by zero growth, no new nuclear power plants would be required. The implied conclusion is that nuclear power is not the ideal solution to the United States' energy problems.

So where do we go from here? There are a half-dozen good answers to our problem. In fact, it is a combination of many technologies that will meet the challenge of supplying future generations with adequate power to operate a highly industrialized society, while at the same time remaining in true harmony with nature. The key is renewable energy sources, sources that are plentiful and continuously replenished by nature. There are many such natural forces, both random and cyclical, that can be harnessed to provide us with all the energy we could ever want. Some could be developed rapidly; all we need is the determination and a healthy concern for the entire society rather than one segment of private enterprise.

There is no lack of energy; it is there for the taking. This is the message the electric wishing well repeats over and over again, with every rise of the tide, with every gust of wind, with every sunrise.

4

Some Dark Horse Energy Systems

The United States is presently facing two contradictory problems: shortage and overabundance. We have already discussed the first problem—a critical long-term shortage of fuel (gas, oil, eventually coal and possibly nuclear fuel). The second is related to the consumption of materials in our affluent society. As a nation, we produce an unbelievable amount of garbage—or, as we shall refer to it here, solid waste. Individuals produce waste, families contribute to the growing heaps of it, businesses add enormous amounts, and farms and industry join in. The result is a huge mountain of solid waste. Public health statistics show that in 1920 the amount of refuse averaged out to 2.8 pounds per person per day. In 1960 it had nearly doubled to 4.2 pounds per person; by 1980 that number may rise to 5.5 pounds per person. Americans annually throw away about 30 million tons of paper, 4 million tons of plastics, 48 billion cans and 26 billion bottles and jars. Hundreds of millions of dollars and thousands of acres of land are presently required to dispose of this enormous amount of solid waste, which is threatening to bury our cities and pollute our waters.

Until recently, we thought of solid waste as something to get rid of, something to throw into incinerators, dump into the sea or bury under the ground. Garbage was merely a nuisance in our lives. But with some careful planning, solid waste can solve a number of our problems.

Solid waste, that loathsome burden in our lives, could be a new source of fuel. How much energy could it provide? If we burned all the solid waste generated by Americans, we could produce energy equivalent to 290 million barrels of oil each year. At $10 per barrel, this would mean a reduction in oil imports of $2.9 billion. We would also realize tremendous savings through reduced land use (for municipal dumps) and maintenance of dump facilities, not to mention the reduced ground and water pollution resulting from chemical damage by giant mounds of garbage. If all this material was properly processed and burned, the resulting solid material would be only 5 percent of its original volume. It would also be much more chemically stable as land fill. The heat value of urban waste is about 50 percent that of coal, and it is very low in pollutants. And here we are not only throwing millions of tons of the stuff away, but suffering it as a serious economic and environmental burden.

Consider a report published in 1972 by the United States Department of the Interior (Circular IC8549). It states that a significant amount of oil could be produced from organic waste. In 1971 the U.S. demand for oil was about 5.8 billion barrels per year. Approximately 1.4 billion barrels were imported. The 1.098 billion barrels of oil that could have been produced from organic waste that year would have reduced the amount of imported oil by 77 percent, an impressive reduction by any standard. The oil produced from the waste could have substituted for the oil used by electric utilities to generate electricity.

The amount of synthetic gas capable of being generated by the solid waste accumulated in the United States amounts to 8.8 trillion cubic feet (TCF). This is a healthy 38 percent of the 22.8 TCF of natural gas used by the United States. Since the amount of solid waste generated in this country is increasing, it will always represent a significant source of fuel, especially for the electric utility industry.

Three characteristics of solid waste make it valuable as an energy source. First, there is a lot of it (solid waste is continuously being generated) and the main concentrations are in the cities, which require a great deal of energy. Second, it is better than cheap: this form of fuel is available at a negative cost—that is, people will actually pay to have it taken away. Third, most solid waste is organic. Between 70 and 80 percent of it is made up of paper, cardboard, wood, plastics and food scraps—organic materials which, when exposed to high temperatures, will break down into gases, liquids and solids containing chemical energy. When burned, each of these materials will produce, besides heat, water and carbon dioxide, which represent the lowest ecological energy level. In plain English, burning garbage does not introduce a lot of destructive chemicals into the atmosphere; it is like burning low-sulphur coal. In fact, 1 ton of solid waste will provide

104 The Electric Wishing Well

about as much energy as 65 gallons of fuel (No. 2) oil or 8000 cubic feet of natural gas. If used as a fuel by the utilities, the nation's refuse could provide about 8 percent of our electricity. And the U.S. refuse supply is not only increasing in volume, but also in caloric (heat) value. It is low in sulphur (0.1 to 0.2 by weight), and can be burned so that it produces less fly ash than pulverized coal.

There is no question that solid waste is valuable as a fuel, but do you just shovel the garbage into a boiler and let it burn? Unfortunately, it is not quite that simple. Solid waste contains a certain amount of nonflammable material. Besides, garbage is not the right size for efficient burning. So a regular process must be followed in order to get the greatest amount of energy out of solid waste. After it has been collected, the waste is brought to a processing plant (which can be on the grounds of a utility, or centrally located if it services a number of utilities). The first step is to shred the refuse in a large machine called a hammermill, which chops the solid waste into small pieces. Once shredded, the burnable portion of the refuse must be separated from the nonburnable portion by introducing the shredded waste into a classifier unit. As its name implies, this machine classifies or separates the waste into its various components. This is done by the use of air (see Figure 27). The waste is fed into a vertical chute, but not all of it falls down the chute. A strong flow of air moving up the chute blows the paper and other light materials upward and they are then collected at the top of the chute. Heavy materials such as metals and glass fall to the bottom and onto a magnetic conveyor belt, which holds the metallic waste and dumps the nonmetallic waste into a storage bin. The nonburnable nonmetallic waste is carted away and used as ground fill. The amount of metal obtained in this process could be considerable. Manufacturers will pay $30 to $60 per ton for scrap steel and $300 for scrap aluminum. Since the average ton of refuse yields about 150 pounds of steel and between 10 and 20 pounds of aluminum, copper, brass and other metals, the income to a utility could really add up. A utility might burn as much as 100 tons per hour; so the scrap metal could bring in about $25,000 per day. The U.S. Environmental Protection Agency estimates that the national volume of refuse for one year would yield about $1 billion in scrap metal.

The burnable waste is carted off by truck to the electric utility, where it is fed into huge steam-boiler furnaces. Here the shredded waste is burned in the furnace along with another fuel (until now only pulverized coal has been burned with refuse, but the method can be adapted to oil- and gas-burning furnaces). It was originally thought that only 5 percent of the total fuel burned by a utility could be solid waste if efficient operation of the furnace was to be maintained. But it turns out that as much as 35 percent of the fuel can be refuse.

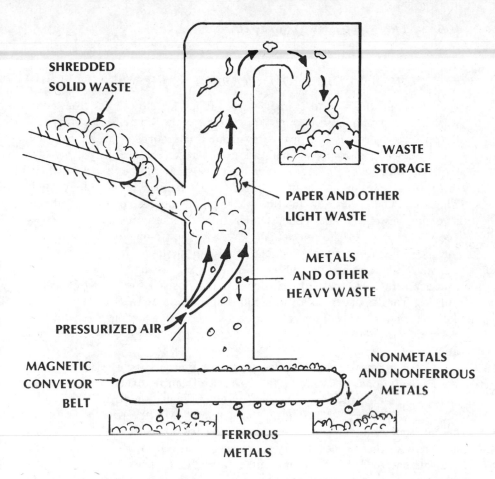

SHREDDED SOLID WASTE

WASTE STORAGE

PAPER AND OTHER LIGHT WASTE

METALS AND OTHER HEAVY WASTE

PRESSURIZED AIR

MAGNETIC CONVEYOR BELT

NONMETALS AND NONFERROUS METALS

FERROUS METALS

FIGURE 27. A classifier system separates solid waste into its various metallic and nonmetallic components.

Naturally, the system is not without problems. Furnace walls and boiler pipes seem to corrode more rapidly when large amounts of refuse are burned. Also, blowers and the moving parts of the conveyors often get jammed. But improved design and altered burning rates and patterns, plus additional research and development, will solve these relatively minor problems. What is important is that there are no insurmountable problems standing in the way of a national network of solid-waste systems to generate electricity. Of course, burning solid waste is not the only method of extracting its energy or using it as a source of fuel.

GAS AND OIL FROM GARBAGE

It is also possible to extract a gas similar to natural gas from solid waste. The method used to do this goes by the name of "pyrolysis." Pyrolysis is a process by which certain materials (organic in nature) are broken down or decomposed with the use of heat. The products of pyrolysis have potential energy (in a chemical form) and can be used as fuel. In this process solid organic waste is placed in a chamber and rapidly heated. The heating takes place in an atmosphere practically free of oxygen; therefore the waste will not ignite. Under high temperatures, the molecules (combined group of atoms) actually explode; the fragments of the exploding molecules form many products, including methane, hydrogen, carbon dioxide, carbon monoxide and water. Further processing of these gases will produce a fuel gas that can be used as an industrial fuel and by electric utilities. The Environmental Protection Agency states that by 1990 about 1.2 quadrillion BTUs of energy will be available from various forms of fuel derived from solid waste. Gas is one of the best forms since it has the least amount of pollutants.

Besides burning solid waste, and extracting gas from refuse, a third path to additional fuel may be taken: fuel oil can be extracted from solid waste. Generally, the same method, pyrolysis, can be used to extract fuel oil from solid waste. And the oil that comes out of the process is low in sulphur and ready for immediate use by industry or electric utilities.

Are all these processes for extracting energy from garbage practical? Has anyone really tried to make them work? The answer is a resounding yes. For years the Europeans have been burning solid waste to generate heat and extracting gas and oil from refuse. In the United States we have been a little late in getting started, but there are a couple of places where energy extraction from solid waste is a big business—Nashville, Tennessee, for example. A huge complex of downtown buildings is being heated and cooled by energy derived from burning urban waste. At two huge incinerator-boilers 1200 tons of solid waste per day are burned. The heat generates about 400,000 pounds of steam per hour, which is used for heating the buildings in winter and for driving a large refrigeration system in the summer. By 1978 this system will save, each year, about 140,000 tons of coal, or 2.4 billion cubic feet of gas or 20 million gallons of No. 2 fuel oil, not to mention the savings in land for waste disposal and operating costs. Perhaps even more important, this solid-waste energy system will drastically reduce pollution. For the building complex it is serving, the system reduces particulate pollution by 372,000 pounds per year, and sulphur dioxide by 359,000 pounds each year. In Baltimore half the city's refuse is being used to produce fuel gas. In San Diego two communities, Escondido and San Marcos, are using their garbage to produce fuel oil by the pyrolysis method. St. Louis probably has the most advanced solid-waste energy plant presently in operation. Here, about 1600 tons of solid waste per day are being processed and burned to help generate electricity. Garbage disposal problems have been solved, pollution problems are being reduced, fuel demands are being met and some money is being made in the process. In fact, the St. Louis project is such an excellent model that visitors come from all over the world to see the system in operation.

Also interested are a lot of big companies such as Babcock and Wilcox, Union Carbide and General Electric. This could turn out to be a really big business because we need fuel, we need to reduce pollution, we need to do something with the tons of garbage that are piling up and we need to reduce our dependency on foreign fuels—and solid-waste energy systems help solve all these problems.

Even if we did use all the solid waste available, however, there would still be a need for a lot more energy. All the solid waste in the United States could not supply the electricity that will be needed by an electricity-oriented society. Other methods must be sought.

ENERGY FROM ROCKET ENGINES

A new technology with the tongue-twisting name of magnetohydrodynamics (or MHD) could provide large amounts of elecricity. In the late 1950s and early 1960s the Avco Everett Research Laboratory was contracted by the government to develop a system that could simulate the conditions of a missile reentering the earth's atmosphere at high speed. Scientists under the direction of Dr. Arthur Kantrowitz developed a rocket engine device that would develop high speed and extremely hot exhaust gases, the conditions a rocket would encounter as it slams into the earth's atmosphere and heats up because of the friction of the air. The research work also gave birth to a system that could produce electricity the way an electrical generator does, except that the MHD unit had absolutely no moving parts. The effort has been continued by Avco and a growing list of other companies and government agencies.

In conventional electrical generators an assembly called the armature rotates in a magnetic field, thereby generating electricity. In MHD the armature is replaced by a gas (see Figure 28). The gas contains material that makes it capable of conducting electricity, so as it passes through the magnetic field, it acts like an armature. As seen in Figure 29, fuel (oil or coal dust) is introduced or injected into a special area called a burner

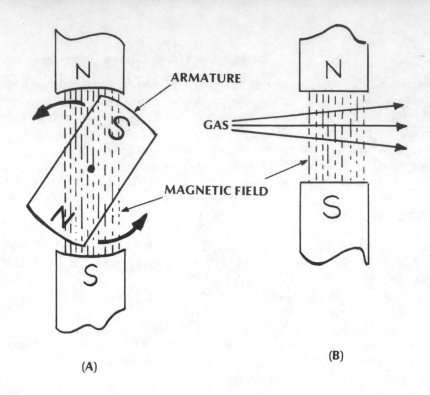

(A)

(B)

FIGURE 28. Conventional method of generating electricity (A). In MHD technology, a gas is used in place of a rotating armature (B).

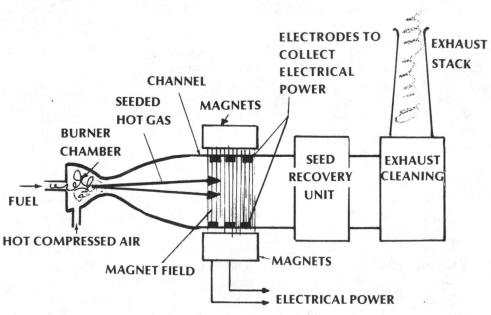

FIGURE 29. MHD open-cycle system.

chamber. Also injected into the chamber is very hot compressed air (called preheated air). The compressed air and fuel are mixed and ignited. But besides the air and fuel, another material is injected in the burner chamber, a material referred to as seed (usually metal particles), which improves the electrical conductivity of the gas. Potassium, cesium or other substances can be used for seeding. The combustion in the burner chamber raises the temperature to about 5000° F., which causes the gas to expand very rapidly and therefore leave the burner chamber at an extremely high velocity. This is similar to the long fiery tail behind a rocket engine, but in the MHD system the exhaust gases remain within an enclosed area called a channel. Remember that the metal particles or seeding travels at high speed with the exhaust gases. Surrounding the channel are a number of powerful magnets, the magnetic fields of which cut through the channel in the path of the hot exhaust gases. The setup is now correct (and similar to a conventional generator) for producing electrical power. The seeded hot gases, which operate like an armature, cut through the magnetic field. Whenever a conductor moves through a magnetic field, electricity is produced; therefore electricity is produced in the channel of the MHD system. In other words, electrons are broken free from the atoms of the hot gas and allowed to flow freely. A number of devices called electrodes are located in the channel. Their job is to collect the electricity and guide it onto an electrical wire, where it is transmitted to the point of use. Once the gas has passed through the magnetic field, it has completed its basic function. But it still contains the seeding material, which is quite valuable and so is recovered in a special unit and returned to the burner chamber. Methods have been developed to remove 99.95 percent of the seeding material from the hot gas. Finally, the exhaust gases are passed through a machine that cleans them of chemicals and particulate matter before allowing them to be vented to the atmosphere.

This system is called an "open-loop system," i.e., it is open at one end to allow the hot exhaust gases to escape to the atmosphere. Actually, the heat of an MHD unit is not all wasted. First of all, part of the exhaust gas is tapped and used to run a small turbine-driven air compressor. This is the compressor that supplies the air to the burner chamber. The main stream of exhaust gas is still very hot when it leaves the MHD unit. In fact, it is powerful enough to drive a gas turbine, which, in turn, could operate a conventional electrical generator. After that, the exhaust gases would still have enough heat energy for use in a low-pressure turbine system that operates on a highly volatile gas such as ammonia. This combination of generating units is called a combined-cycle system (discussed in Chapter 1).

Heat can be used very efficiently by an MHD system, which is one of its main advantages over conventional systems. Today's best atomic power plants operate at only 32 percent efficiency, and conventional steam generators have not gotten past 40 percent. MHD systems can produce electricity with 60 percent efficiency when they operate in large units. And there are other advantages.

MHD systems have no moving parts except for a few small accessories such as the air compressor. With no moving parts there are no energy losses or wear through friction. Furthermore, when used in a combined cycle with a gas turbine system, MHD requires practically no cooling water. A 2000-MW system could be operated in the middle of the desert. Another advantage is fuel flexibility. MHD units can operate on almost any fuel—gas, oil, coal dust, hydrogen and oxygen, shale oil, whatever is available. Unlike large steam generator systems, an MHD unit can be turned on almost instantly, so it can be used as either a base load or a peaking system. And there is an indirect advantage—since an MHD unit is much more efficient than other systems, it uses a lot less fuel and generates more electricity for each unit of fuel. The result is less pollution. In fact, the amount of sulphur dioxide, nitric oxides and other harmful pollutants from the MHD is about 75 percent below the standards set by the Environmental Protection Agency.

Although the greatest amount of research work has been done in the open-cycle MHD system, a number of companies are beginning to think more seriously about a more advanced version of MHD—the closed-cycle system. As its name implies, in this type of MHD system exhaust gases are not vented to the atmosphere. Fuel is not burned and passed through the channel assembly, as in the open-cycle approach. An inert gas such as helium, argon or neon, which is seeded with a metallic substance, is used as the working fluid. As shown in Figure 30, the gas is injected into a chamber similar to that of an open-cycle system. However, here the gas is indirectly heated from another source. This can be a coal fire, oil or natural gas. (Some scientists suggest a nuclear reactor.) As the gas (containing seeding material) is heated, it expands and travels at high speed through the magnetic field in the channel assembly of the MHD generator, thereby generating electrical power. When it leaves the channel assembly, the hot inert gas passes into a large cooling chamber where it is allowed to condense and cool. A pump is then used to transfer the gas back to the gas heating chamber where the process starts all over again.

One advantage of this system is that it operates at a temperature about 2000° F. lower than an open system. Therefore the possibility of breakdown because of extremely high temperatures is reduced. Which brings us to the reason why MHD has not advanced as rapidly as one would wish. Materials must be found that can withstand heat up to 5000°

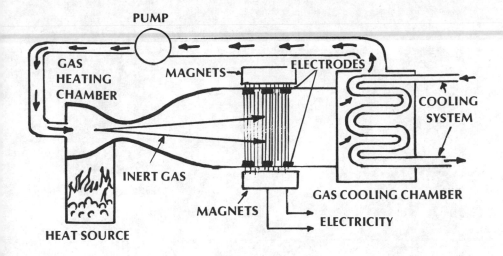

PUMP

GAS HEATING CHAMBER

MAGNETS

ELECTRODES

COOLING SYSTEM

INERT GAS

MAGNETS

GAS COOLING CHAMBER

ELECTRICITY

HEAT SOURCE

FIGURE 30. MHD closed-cycle system.

F. continuously and last for thousands of hours of operation. When coal dust is used as a fuel, for example, a hot liquid called slag is formed at 5000° F. Coal slag is so corrosive that it will destroy almost any material it contacts, including ceramic bricks. Obviously, a great deal of research on heat-resistant materials will have to be performed before MHD is practical.

Within the channel assembly short circuits may develop between electrodes (the parts that collect the electricity) because of the passage of high-speed, hot, abrasive gas over the electrodes. Another problem is condensation and penetration of seed material, which causes arching between electrodes. Then there are magnet assembly problems. MHD generators require much more powerful magnets than conventional electrical generators because the seeded gases that act as the armature do not conduct electrical current as well as copper. Superconducting magnets will probably be used in MHD generators. However, a lot more research must be done to develop the practical large magnets that are necessary for large MHD power plants.

None of these problems is a major stumbling block in the way of MHD power. Dr. William D. Jackson, principal research scientist for Avco's MHD program, does "not believe that there are any unsolved materials or engineering problems." Materials have been selected and designs have been completed; what is needed is further support from the government and electrical utilities. Throughout the 1960s Avco Everett Corporation was doing the major share of the work in MHD research, but because of limited funds, progress was slow. The first working unit, the Mark 1, was built in 1959; this experimental model was capable of producing

Technician connects electrode leads to MHD generator channel. In the future, such units will provide large amounts of high-efficiency, clean electrical energy. (Courtesy Avco Everett Research Laboratory)

10,000 KW (kilowatts) of power for only 10 seconds. By 1965 the Mark 5 was producing 32,000 KW for a full 60 seconds. The short operating time reflects the problems of running an MHD at high temperatures. In its latest work Avco has concentrated on endurance. Its Mark 6 generator can produce 250 KW for 100 hours.

The Westinghouse Corporation, General Electric, the Bureau of Mines, the Office of Coal Research, the U.S. Air Force and a number of universities are all involved in developing this very promising technology. Of course, we are not the only nation interested in MHD. Japan, completely dependent on foreign oil, is seeking a highly efficient method for using oil so it can reduce its imports substantially. Germany, which has a good supply of coal, is also doing a great deal of research work in MHD. But the country way out front in MHD development is the Soviet Union, which has already built and is operating a large pilot plant capable of generating 75,000 KW of electrical power. Operating on natural gas (fairly abundant in Russia), this plant, designated the U-25, works in combination with a steam turbine system and feeds electricity into the Moscow power grid.

Scientists at General Electric believe that a 100,000-KW demonstration plant could be in operation as soon as 1980 (well ahead of the breeder reactor program). Scientists at Avco are also optimistic about MHD, maintaining that only $8 million per year is needed to develop a pilot plant capable of operating for 100 hours at a power level of 50,000 KW, which could begin preliminary operations by the end of this decade. Avco's program would lead to the construction of a demonstration plant, whose design would be based on experience derived from long-duration tests of the pilot plant. The demonstration plant, which would cost about $336 million, would be capable of generating up to 1,000,000 KW of electricity and would be producing power by 1984. It is not overly optimistic to assume a full-scale MHD power plant could be feeding 2,000,000 KW of electrical power into a U.S. power grid by 1990.

MHD has a great deal to offer: high efficiency, reduced air pollution, no water pollution, safe operation and extreme flexibility in fuel requirements. It is an excellent alternative to atomic power—or other electrical power system that will quickly deplete our supply of fossil materials.

ELECTRICITY FROM WATER

As early as 1958, before the United States made its famous commitment to land a man on the moon before the end of the decade, the Pratt and Whitney Company was hard at work on a relatively new approach to generating electrical power. By January 1962 it was conducting initial tests on an experimental unit (PC2), the first complete automatically operated power plant of its kind. Two months later Pratt and Whitney was contracted by the North American Aviation Corporation to develop a dependable fuel cell that could supply 1500 watts of electrical power to the Apollo command and service module. Pratt and Whitney had the foresight to recognize the long-range value of this device: in May 1962 it was operating a 500-watt fuel cell for Columbia Gas Systems at a Stanton, Kentucky, pumping station. This was the beginning of commercial fuel cell development. Four years later P & W delivered a 3750-watt fuel cell that operated on natural gas. Meanwhile, the fuel cells in the moon program were logging thousands of hours. "Fuel cell" became the "in" term in much of the scientific community during the mid-1960s.

Why all the excitement over this new technology? Simply because fuel cells potentially offer advantages and characteristics that cannot be matched by any other method of producing electrical power. Fuel cells are the only energy conversion method (i.e., producing electricity from fuel) that does not depend on a heat cycle of any kind. Therefore it can offer very high efficiencies; fuel cells have already been operated at 70 percent efficiency, more than twice that of atomic power plants. Also, fuel cells can operate on natural gas, oil, coal dust, derivatives of petroleum, methane and, best of all, hydrogen and oxygen, which can be obtained from water.

Their advantages go on—small size, for example. A fuel cell power plant capable of supplying electricity to 20,000 people measures only 18 feet in height and would sit on less than half an acre of land. Furthermore, fuel cells are extremely quiet, have no moving parts, generate practically no heat (you could place your hand in the exhaust of a fuel cell and not feel uncomfortable) and produce no air or water pollution. They are perfectly safe under just about all conditions; a large fuel cell power plant could be situated in the center of a heavily populated area. A fuel cell power plant is also extremely simple in operation, fully automatic and requires minimum personnel in attendance. Its fuel can be supplied through underground pipes, so there is no need for heavy truck or train traffic for deliveries, nor for an ugly stockpile of coal on site. Because the efficiency of fuel cells does not depend on plant size (whereas atomic plants and MHDs must be large to be efficient), moderate-sized power plants could be located in populated areas where the demand for electricity is greatest, thereby minimizing energy loss from long-line transmission. For that matter, large apartment or office buildings could have their own power plant right on site. Finally, because large fuel cell power plants can be built of smaller modular units, it is much easier to match power plant output to demand— that is, one can operate 10 percent, 60 percent or 100 percent of the plant without long starting up procedures or loss of efficiency. This is something that cannot be done with large steam turbine generating plants. What is the secret of these electrical power geniis?

The operation of a fuel cell is quite simple. Basically, it is like an ordinary battery with a few added features. A fuel cell system is made up of three sections (see Figure 31): the re-former, the fuel cell and the

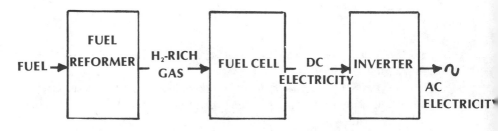

FIGURE 31. A fuel cell system produces electricity from various types of fuel without requiring heat.

inverter. The re-former changes or re-forms the fuel that is used (either natural gas, oil or coal products) to a gas that is very rich in hydrogen atoms. In the re-former a small amount of heat is used to generate the hydrogen-rich gas (if pure hydrogen is used, neither the heat nor the re-former is required). The gas that is produced by the re-former then flows into the fuel cell, where the electricity is actually generated. Here the operation becomes a little tricky. As shown in Figure 32, hydrogen-rich gas enters the fuel cell from one side, while oxygen in the form of air enters the cell from the other side. Located in the cell are two electrodes, one for the fuel (hydrogen) and one for the air (oxygen). As the oxygen enters the chamber, it reacts with the air electrode and picks up two electrons. Then it passes through a liquid in the center of the cell, called an electrolyte (which is required to allow the entire chemical action to take place), and reaches the area around the fuel electrode. At this point two things happen. The oxygen combines with the hydrogen to form water (H_2O), and the oxygen atoms deposit their extra electrons onto the fuel electrode (shown as negative charges in Figure 32). So the fuel electrode acquires a lot of extra electrons (meanwhile, the air electrode has lost a lot of electrons).

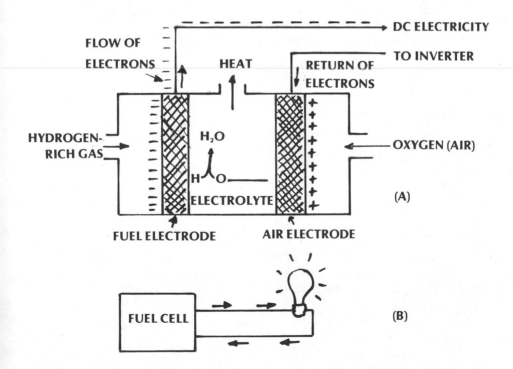

FIGURE 32. A fuel cell combines oxygen and hydrogen to generate electricity.

The wires that are connected to both electrodes have electricity on them. Electricity is the flow of electrons through a wire; here we have a lot of electrons, so we have electricity. When the electrons have flowed through the wire and performed some work, such as lighting a bulb, they flow back to the fuel cell and return to the air electrode, where they belong. This entire process continues as long as fuel and air are supplied to the cell. What a fuel cell does in one step (without any moving parts) it takes a conventional power plant three steps to accomplish (generate steam, drive a steam turbine, rotate an electrical generator). But since the electricity produced by a fuel cell is direct current (DC) and most U.S. homes operate on alternating current (AC), a third unit, an inverter, is needed to change the DC power from the fuel cell into AC power. Inverters operate at a very high level of efficiency—better than 90 percent.

It was mentioned that fuel cells reduce pollution; actually, they virtually eliminate these unwanted by-products of energy generation. For example, fuel cells produce about 1000 times less sulphur dioxide than do fossil fuel steam power plants, and about 16 times less oxides of nitrogen, 12 times less hydrocarbons and about 5000 times less particulates. Yet, despite all these advantages, the government has provided no significant support for the development of commercial fuel cell power plants, though it has sunk billions of dollars into the problem-laden, hazardous breeder reactor. Perhaps it is still doubted that commercial fuel cell operation is practical.

In 1970 Pratt and Whitney began testing commercial fuel cell power plant units (designated PC 11) in 17 cities, including New York, Chicago, Los Angeles, Boston, Cleveland and Detroit. It wanted to prove that fuel cells can operate in any environment. The power plants logged almost 100,000 hours with no serious problems. A PC 11 (12,500-watt) fuel cell installed in a home in Wetherfield, Connecticut, has thus far accumulated 200,000 hours of successful operation. Unquestionably, fuel cell technology can contribute significantly to satisfying our demands for electricity without endangering our environment.

In January 1967 a group of 28 gas utilities joined in a cooperative program with Pratt and Whitney to develop gas-operated commercial fuel cells. Among the companies were New York's Consolidated Edison, Boston Edison, Consumer's Power Company of Michigan, Niagara Mohawk of New York, Southern California Edison and Philadelphia Electric. A lot of engineering design work and materials research must be done in order to make large fuel cell power plants more economical to construct and key parts such as fuel cell electrodes last longer. But many companies are now researching these areas—General Electric, for example, which developed a fuel cell power unit for the Gemini space capsule in the early 1960s. The

This fuel cell power plant, which can generate 26 million watts (26-MW) of electricity, is capable of supporting the electrical demands of a community of 20,000 people. (Courtesy Pratt and Whitney Aircraft)

Exxon Corporation is doing work in this area in a team effort with France's Alsthom and Company (which manufactures Peugeot autos). Westinghouse, Shell Oil and a host of others are busy on this promising technology. Meanwhile, General Motors is conducting research to develop a fuel cell that could provide the electrical power to run an automobile. It would have a great advantage over battery cars: no recharging would be required and the power plant would weigh a mere fraction of its equivalent in batteries. And commercial fuel cells for electrical power generation won't be very long in coming: a 26-MW unit could be available by 1978 or 1979. It is not unreasonable to expect that by the mid-1980s there will be a significant number of 500-MW power plants operating around the country. They have a lot to offer and certainly deserve full support from government and industry.

One very important point to consider regarding fuel cells is that they operate best on pure oxygen and hydrogen. With these fuels, a fuel cell will run virtually pollution-free and at an efficiency rate approaching 80 percent under the right conditions. Certainly this would be an ideal objective—a quiet, safe, efficient, pollution-free electrical generating system that operates on a fuel that is virtually unlimited in supply (water). There are some questions, of course. Can we generate hydrogen and oxygen economically? Is hydrogen safe to handle? Can it be safely stored? How else can we use it? Can hydrogen solve our fuel shortage problems? How does hydrogen compare with other fuels? Is it the ultimate fuel? Is the concept of a hydrogen society realistic? It is to these questions that the following chapter is addressed.

5

From Water to Water

The energy crisis has made us painfully aware of the limitations of fossil fuels. What we need is a substitute fuel that is clean and virtually unlimited in supply. Where can such a fuel source be found? All around us! We drink it; we bathe in it; it rains on us. The greatest source of clean fuel is water. Why? Because water (H_2O) contains hydrogen (H) and oxygen (O)—the two substances that fueled the upper stage of the Apollo spacecraft. Hydrogen, the simplest of all elements, is the ideal and ultimate fuel. It is storable, cheaply transportable, abundant, efficient, burns with no adverse effects on our environment and works in perfect harmony with nature.

Hydrogen is the only fuel that can be consumed in quantities sufficient to generate all the energy we need without destroying the environment. When hydrogen burns (under most conditions), its only combustion product is water. None of the traditional fossil fuel pollutants such as carbon monoxide, carbon dioxide, sulphur dioxide, hydrocarbons, particulates or photochemical oxidants will be produced with a hydrogen flame. In the fossil fuel cycle a heavy burden is placed on nature. It takes nature millions of years to produce fossil materials (such as coal, oil and gas); once burned as fuels, the pollutants from combustion are returned to nature and may require millions of years to be reconverted to fossil fuels. This means that nature must carry a double load, i.e., produce the fuel and clean up after us. But in the hydrogen cycle the raw material (water) is available in large quantities and the only products of combustion are water and heat. The water is immediately returnable to the environment as one of its natural elements. Therefore, except for thermal pollution, hydrogen fuel can be used in unlimited amounts with no adverse ecological effects.

Another reason hydrogen is the ideal fuel is the trend to electricity as a basic form of energy. Our use of electricity is not constant or uniform; consumption varies with the time of day and year. But most types of conventional power plants achieve their best efficiency when they can operate at a constant rate. Nuclear power generating stations, for instance, operate most efficiently at a constant *maximum* power output. But since we do not need all that power constantly, the extra electricity must somehow be stored. Other proposed new energy forms such as solar power, wind power and tidal power, though they are valuable sources of large quantities of energy, suffer the disadvantage of being inconstant in availability. All that is needed to make them practical sources of energy is a method for storage. Hydrogen is a good method of storing electrical energy—a fact that will be of utmost importance when the nation comes to rely on electricity for more than 50 percent of its energy.

Finally, hydrogen can be our solution to the problem of long-range energy transmission. In the future much electricity may be produced at stations located in remote areas (for example, solar farms in the desert, tidal power plants on the ocean, wind-powered generators on the great lakes). But up to 10 percent of electrical energy can be lost in transmission over long-distance power cables. One way to reduce this loss is to transport electricity in the form of hydrogen gas (produced from electrical power) through long-distance underground pipes. At the electrical power plant water would be converted into hydrogen by a process called electrolysis. The hydrogen gas would be pumped through pipes to distant cities or industrial locations, where it would be used to run fuel cells or MHD systems. Such a system would preclude the construction of thousands of miles of high-voltage transmission towers or underground cables, which are extremely expensive. Therefore hydrogen would make remote power plants more practical and efficient.

Since hydrogen presents such excellent credentials—it is a good fossil fuel substitute, it is available in quantity, it is nonpolluting, it is a fine energy storage medium and a good medium for energy transmission—it is inevitable that it will be used by our society as a basic fuel supply. Why, then, are we not already using this marvelous fuel?

Price, for one thing. With our present methods of production, hydrogen costs more than other types of fuel. The economics of producing and using hydrogen are complex. Right now hydrogen cannot compete with natural gas or gasoline as a fuel; because it is so light, it cannot produce as much heat value on a volume basis, which makes it more expensive to use.

But the picture will soon change. One reason for the high cost of hydrogen is the expense of the electricity used to produce it. Eventually, however, huge power plants of the 30,000-MW size may produce electricity much more cheaply, which will bring down the cost of producing hydrogen. In addition, as continental oil and gas begin to run out, their prices will rise sharply. Our entire supply of economically recoverable petroleum will be consumed in less than 60 years. U.S. production of oil has already passed its peak and is on the decline; drilling of new wells will supply barely enough to maintain present levels of production, while demand continues to grow. Surely this means much higher petroleum prices. The cost of mining coal is rising, and hence coal is becoming more expensive. Hydrogen, then, will be not only more cost competitive, but also one of the few fuels left to man.

The sooner we change over to a hydrogen economy, the better off we will be. Hydrogen produces a tremendous amount of heat per unit weight. Compare it with other future fuel candidates. Combustion of ammonia produces 8000 BTUs of heat per pound; ethyl alcohol, 12,800 BTUs; JP (jet fuel), 18,000 BTUs; and one of the best fuels, methane, 21,120 BTUs. Hydrogen produces 51,590 BTUs per pound. Since it burns without noxious fumes, it can be used in unvented areas. With a hydrogen fuel system, it would be possible to build a home heating furnace without a flue, saving the cost of a chimney and increasing the efficiency of a home heating system by up to 30 percent. The water resulting from the combustion of hydrogen would provide a home with sufficient moisture for maximum comfort. Our present systems and equipment can be easily adapted to hydrogen, so we could continue to use gas systems (stoves, boilers, refrigeration, etc.) already in operation. In fact, every existing fuel-consuming device, as well as factories and electrical power plants, can be converted to use hydrogen gas. Such conversions would eliminate factories and electrical power plants as significant sources of pollution. Has anyone given serious consideration to these advantages? Are any concerted efforts being made to bring about the hydrogen economy?

There is some activity in this area, but not enough. The idea of using hydrogen as a fuel is certainly not new. As far back as 1933 Rudolf A. Erren, a German inventor working in England, suggested the large-scale production of hydrogen from extra electricity available during off-peak hours. Erren also did pioneering work on converting the internal combustion engine to operate on hydrogen. During the same period an Englishman named F. T. Bacon, who did early work in fuel cells, recommended hydrogen as an excellent medium for storing energy. But real advances in technology had to wait until the 1960s when the National Aeronautics and Space Administration invested millions of dollars to find the best methods of producing hydrogen, liquefying it, handling it, burning it and storing large amounts of it in liquid form. Fifteen years of successful use of liquid hydrogen has proved that it can be readily adapted as a basic fuel. But a lot more developmental work must be done before this fuel will be avilable to the general public.

In March 1974 the University of Miami held a conference on the use of hydrogen, hoping for a modest attendance by the scientific and engineering community. To the organizers' surprise, 750 people showed up eager to get the latest information on this promising technology. Similar conferences are now popping up all over the country. The gas industry itself is interested in hydrogen technology, since it is expected that natural gas will be the first fossil fuel to be depleted. A group of researchers led by Dr. Derek P. Gregory of the Institute of Gas Technology in Chicago had the foresight to begin work on hydrogen almost 15 years ago on a tiny research budget. They now have significant financial support from the American Gas Association. Also involved in hydrogen research are large companies such as Jet Propulsion Laboratory, General Electric and General Atomic. Only a modest amount of research and development is needed to prepare our society for transition to a hydrogen economy. Once hydrogen is available in large volume, we will no longer be dependent on fuels whose supply is subject to the ever-changing winds of the political arena, nor will we be in danger of disrupting our own ecological system. But where is all this hydrogen going to come from?

PRODUCING HYDROGEN

Because of the great need for hydrogen by such industries as steel, chemicals and petroleum refineries, several processes for producing this gas have been in operation for nearly two decades. Right now one of the cheapest methods of producing hydrogen is to extract it from natural gas. In one process called partial oxidation, natural gas (or another hydrocarbon feed) is preheated and fed into a large tank. Also fed into the tank are high-pressure steam and oxygen. The temperature within the tank is around 2500° F. These three elements (steam, oxygen and natural gas) react with each other under very carefully controlled pressure and flow-rate conditions. The result is the generation of hydrogen gas (see Figure 33), which is passed through a cooling stage and then through a series of steps to remove unwanted chemicals such as carbon monoxide (CO) and carbon dioxide (CO_2). The hydrogen, which has a purity of about 96.5 percent, is then pressurized to make it ready for use.

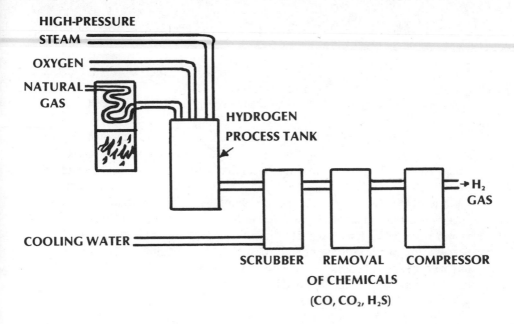

FIGURE 33. Simplified diagram of partial oxidation system used to produce hydrogen.

In another method called catalytic steam re-forming a mixture of steam and hydrocarbon material (natural gas, oil, etc.) is heated to at least 1100° F., and introduced into a special furnace with a large number of metal tubes. The tubes contain a material called a catalyst. The catalyst causes the steam and the hydrocarbon material to chemically react, thereby releasing hydrogen gas. Carbon products are then removed from the hydrogen, which is pressurized to make it ready for use.

The problem with both of these methods is that they are based on fossil fuels. A hydrogen economy would require well in excess of 100 TCF of H_2 gas per year, which would soon exhaust our fossil material supply, not to mention the pollution problems encountered in producing that much gas. The best route to hydrogen energy is using a source of material that is both plentiful and nonpolluting—water.

Water is comprised of two elements, which, when combined and ignited, burn with a furious hot flame. The water molecule (H_2O) is made up of atoms of hydrogen and oxygen. Hydrogen is the simplest of all atoms, having only one proton ($+$) and one electron ($-$) in its structure. The more complex oxygen atom has eight electrons, revolving around a nucleus (core) that contains eight positively charged protons ($+$) and eight neutrons, which carry no charge. Water is formed when two atoms of hydrogen combine with one atom of oxygen. Since water is in such abundant supply, it is the best candidate for a hydrogen source. But how does one go about extracting hydrogen from water? There are two methods for accomplishing this: the thermochemical process and the electrolysis process.

As its name implies, the thermochemical process uses heat and chemicals to cause water to break down into hydrogen and oxygen. Actually, water can be broken down or dissociated using only heat. If water is heated to about 4532° F., its oxygen and hydrogen atoms will begin to separate. However, there are two problems with this approach. First, the energy involved in heating large quantities of water to this temperature would be so great that the cost of the hydrogen would be extremely high. Second, this method does not allow for easy separation of the hydrogen and oxygen for purposes of collection. So for the time being, direct thermal breakdown of water can be ruled out as a practical method. However, if special chemicals are added to the water, the situation changes radically.

Similar to the petroleum cracking process (in which gasoline and other petrochemicals are produced from raw petroleum), the thermochemical process (water splitting), invented and developed by Cesare Marchetti at the Euratom Laboratories in Italy, takes advantage of the reactions between the added chemicals and the superheated water. In a towerlike tank water heated to about 1400° F. is injected along with such chemicals as calcium bromide and mercury. Under specially controlled conditions, this combination causes the hydrogen and the oxygen in the water to separate. Theoretically, there are many combinations of chemicals that will cause water to break down into its basic elements at even lower temperatures, but a good deal of research and development will be required to make these methods practical on a large scale. The most promising method of splitting water remains electrolysis.

Electrolysis requires only water at room temperature, a small amount of acid and electricity and some very simple equipment. As seen in Figure 34, a large tank is supplied with water, which is mixed with a small amount of acid (or other chemical) that enables the water to conduct an electrical current. This mixture is called an electrolyte and is basically the same mixture found in a car battery. Also located in the tank are two metal rods called electrodes; a source of electricity (DC) is connected to these electrodes. The electrode that is positively charged is called the anode, the negatively charged electrode is the cathode.

Because of a chemical phenomenon, some of the hydrogen atoms of the water loose their electrons and "float" around as positively charged atoms or ions. The oxygen atoms gain extra electrons and therefore become negatively charged ions. Because opposite charges attract each

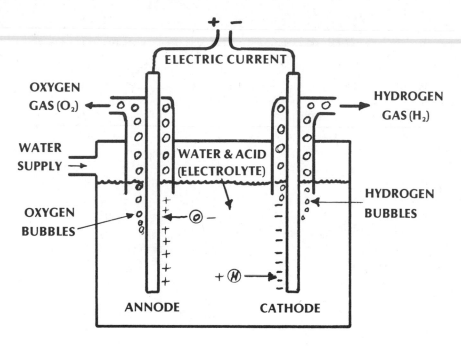

FIGURE 34. Large amounts of hydrogen can be supplied by the electroysis process.

other, the hydrogen ions that are located near the cathode (negative electrode) are attracted to it. Upon contact with this electrode, the hydrogen ion receives an electron and is transformed into a normal hydrogen atom (with no charge). The hydrogen atoms then join in pairs to form hydrogen gas (H_2) and float as bubbles to the top of the cathode and out of the tank, where the gas is collected and compressed.

Meanwhile, at the other end, the oxygen ions are attracted to the positively charged anode. Upon contact, the oxygen ions release electrons and become oxygen gas (O_2), which forms bubbles that float to the top of the electrode and out through the oxygen-collecting pipe, where the gas is compressed. In this way a supply of oxygen and hydrogen is produced from water. The acid in the water does not get used up in the process.

There is no limit to the basic material for this process—all the oceans, rivers, lakes and streams of the world are sources of the fuel. Neither would the quantity of water be diminished, since the burning of hydrogen and oxygen would change these elements back to water. It would be a cycle perfectly in tune with nature: dissociate water into hydrogen and oxygen; burn the hydrogen and oxygen to generate energy; end up with water from the combustion of hydrogen and oxygen. After a few decades of using this system for energy generation, the earth could return to its unpolluted state.

While there is an unlimited supply of water for electrolysis, however, the other major input—electricity—is neither unlimited nor inexpensive. Therefore electrolysis, while offering great advantages in terms of pollution, suffers from the disadvantage of high cost. Since electrolysis relies on electricity, it must reflect the cost of that energy source. Therefore the resulting hydrogen must always cost more than the electricity required to produce it. Fortunately, some of this cost could be offset by the sale of the oxygen that is produced as a by-product—huge amounts of oxygen could be used in coal gasification plants (which would produce synthetic fuel gas), and oxygen could also be used to depollute large bodies of water. In fact, there are literally hundreds of uses for oxygen. But the sale of oxygen would not fully offset the added cost of producing hydrogen by electrolysis.

The essential problem is that electricity reflects the high cost of fuel—mainly petroleum. As long as fuel prices remain high, electricity produced from these fuels will be costly. But other factors will eventually change the picture. Large atomic plants (2000 MW and up) will have to operate at maximum levels to remain efficient. During off-peak hours the excess electrical capacity can be used to produce hydrogen via the electrolysis process.

Furthermore, when fusion plants eventually come into operation, they may have to be built as very large facilities in order to be practical. Electrical power plants generating 50,000 MW might be in operation shortly after the turn of the century. Present or even foreseeable transmission lines might not be able to handle such a huge load; in any case there will be a large transmission loss. Instead, the electrical power could be partially used to produce hydrogen. The hydrogen could be used to fuel small electrical power plants that might be located in urban centers (fuel cells, for example). This method would not incur a great loss; a well-designed electrolysis system could eventually operate at better than 90 percent efficiency. Furthermore, in the future the energy produced by solar, wind and tidal power will at times have to be stored for use at peak periods; hydrogen will be excellent for this use. And unlike other storage methods (flywheels, batteries, heat), hydrogen is mobile. It is a storage medium that permits the transfer of energy to other locations with minimum loss. It is this ability that can make hydrogen economical to use. Billions of pounds of it are presently being used for refining petroleum. Most of this hydrogen is produced from methods that use hydrocarbon materials (gas, oil and coal). But these methods could never supply the necessary amounts of hydrogen if it became the basic fuel for our society. Only electrolysis systems based on huge electrical generating plants could supply the required amount, and the amount would be tremendous.

The equivalent amount of hydrogen gas required to replace the present annual natural gas consumption (22.5 TCF) would be about 70 TCF.

To produce this much hydrogen gas using electrolysis would require about 1 million MW of electrical power. The entire electrical production capacity of the United States is presently a little over 400,000 MW. So we would have to more than double our electrical production capacity just to provide an amount of hydrogen equivalent to today's natural gas consumption. Of course, producing hydrogen is only part of the task. Methods for efficiently storing it must be developed.

KEEPING THE LID ON HYDROGEN

There are three basic methods of handling hydrogen: in pressurized tanks, in liquid form and in special compounds. Hydrogen can be stored in heavy steel tanks that will hold the gas under pressure (at about 2000 PSI) so that it is readily available for use at the twist of a valve. Unfortunately, only a small amount of gas can be stored in a tank and the tank is extremely expensive, rugged and heavy (the storage tank required to contain only 1 pound of hydrogen under a pressure of 2000 PSI would itself weigh about 200 pounds). Therefore this method is only useful for storing small amounts of hydrogen for laboratories or other applications. A similar storage method involves pumping hydrogen gas into a large inflatable tank located far below the surface of the water. The tremendous pressure of the water would support the walls of the tank, allowing it to contain large amounts of hydrogen. But maintaining a large network of these tanks would be difficult and costly, and the source of hydrogen would have to be located nearby, though such a method would be practical for a solar electric plant located out at sea.

The second way to store hydrogen is in large caverns. The gas industry presently stores large quantities of gas, for meeting peak demands, in underground porous-rock formations such as depleted gas fields. Empty mine shafts, salt caverns and other formations could hold significant amounts of hydrogen gas. In 1970 gas companies had a storage capacity (in 337 locations) of 5.2 TCF, or about 22 percent of the annual production of gas. This type of storage would certainly be usable for hydrogen gas. But no matter what method is used, hydrogen in gaseous form takes up a lot of space. To make storage practical, hydrogen must be changed into a more compact form.

Liquid hydrogen is one good answer to the space problem. Hydrogen becomes a liquid at temperatures below $-422°$ F. ($20.5°$ K.), and in liquid form it only takes a fraction of the space it takes as a gas. Unfortunately, because of its extremely low temperature, liquid hydrogen has a tendency to boil away. Although superinsulations have been developed to reduce hydrogen loss to less than 1 percent per day, extended storage of liquid hydrogen would require a tank equipped with a refrigeration system, and this would add significantly to the cost of the storage system because of the equipment and energy required for its operation. But liquid hydrogen enjoys the advantage of being completely mobile and can be stored anywhere. It also has a fantastic capacity for storing energy—pound for pound, hydrogen provides more energy than any other fuel. For example, the world's largest pumped storage system (an artificial lake) in Ludington, Michigan, has an electrical energy storage capacity of 15 million kilowatt-hours. The liquid hydrogen tank at John F. Kennedy Space Center, which holds 900,000 gallons of liquid hydrogen and stands on less than one acre, contains the equivalent energy of 11 million kwh, 73 percent of the pumped storage facility. Many energy storage problems are presently being solved using liquid natural gas (LNG); there are no technical reasons why liquid hydrogen (LH$_2$) cannot perform similar functions.

The third form of hydrogen storage is the metal-hydride system. Hydrogen gas is pumped into a container filled with a special combination of metals called metal hydrides (see Figure 35). Iron and titanium alloy is one combination; magnesium can also be used as a hydride material. The system works as follows: As the hydrogen is pumped into the container, the hydride acts as a sponge absorbing the hydrogen. If the hydride material is iron-titanium, the action of absorbing the hydrogen creates heat. (This is called an exothermic reaction.) The heat can be directed to cooling fins and a cooling water system can be used to carry it off. Metal hydrides are capable of absorbing a great amount of hydrogen. In fact, they can absorb hydrogen to a density three-quarters of liquid hydrogen without having to be cooled to extremely low temperatures. One advantage of using iron-titanium is that the hydrogen absorption occurs at room temperature. (Some metal hydrides, such as magnesium, must be heated to about 650° F. before absorption will take place.) Once the hydrogen has been absorbed, it can be stored almost indefinitely. The hydrogen can also be extracted at any time; this is accomplished by sending electric power through a heater in the container. The heat is absorbed by the metal hydride, which, in turn, releases the hydrogen gas. This system offers another important advantage besides its ability to store hydrogen in large quantities for extended periods of time—safety. No high pressure is required; no superlow temperature is involved. Even if the container were to be damaged or broken open, none of the hydrogen gas would escape. The heat required to release the hydrogen gas from the hydride could also be supplied from hot water pumped through the storage container.

It has been estimated that a 26-MW electrical power plant with 260,000-kwh storage capacity could fit on an acre of land (using fuel cells

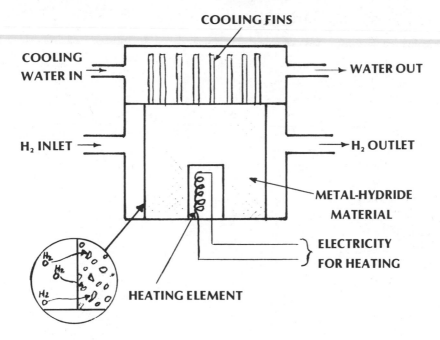

COOLING FINS

COOLING
WATER IN

WATER OUT

H₂ INLET

H₂ OUTLET

METAL-HYDRIDE
MATERIAL

ELECTRICITY
FOR HEATING

HEATING ELEMENT

FIGURE 35. A simplified sketch of a metal-hydride hydrogen storage system.

as the electrical generators). Because fuel cells produce direct current, which is the kind required by hydrogen electrolysis, they would match well with a metal-hydride storage system. And this type of storage system could be safely located in congested urban areas where electrical power consumption is highest. In fact, the first all-hydrogen energy system, including metal-hydride storage, is already under full development by the Public Service Electric and Gas testing laboratory. It is small (12.5 kilowatts), but it could lead to a revolutionary energy system for producing and storing electricity. Storage is the real key to economical electricity, because if we can store large amounts of power for a long time, we can generate electricity continuously at peak capacity, which will allow power plants to run efficiently. With proper storage, we will also be able to use nonconventional, nonconsistent, remote sources of energy.

If hydrogen is to replace natural gas and become the basic fuel for the United States, it will have to be supplied to millions of homes, commercial buildings and factories. It would be completely impractical to deliver the gas in compressed form in heavy steel tanks. Neither would it be economical to distribute hydrogen in liquid form. The clear need is for a practical, dependable, and economical system for mass distribution of hydrogen gas. Such a delivery system now exists: the huge network of natural gas pipelines. Well over a quarter of a million miles of these pipes carry natural gas to most industries and over 80 percent of the homes in this country. This gas transport system could easily be used for gaseous hydrogen. However, because hydrogen is a much lighter gas, its heating value per cubic foot of volume is only about one-third that of natural gas. Therefore the system would have to operate at a higher pressure so that more hydrogen (three times more than natural gas) could be moved through the pipes. These pipes can easily handle hydrogen—in fact, in a number of areas industry is transporting gaseous hydrogen using present pipelines. There is no reason why, with some minor modifications, a great deal of hydrogen could not be economically piped over long distances.

A standard 36-inch gas pipeline has an energy-carrying capacity equivalent to 11,000 MW of electrical power, ten times as much as a single-circuit 500 kilovolt, high-voltage overhead transmission line. It costs roughly 5 cents to transmit the equivalent of 1 million watt-hours over a distance of 100 miles with piped gas. Compare this with 15 cents for overhead high-voltage transmission cables. And underground cable systems can cost up to 40 times as much as overhead lines to construct. So, a hydrogen gas pipeline system is economical, though not necessarily inexpensive. According to Dr. Derek P. Gregory of the Institute of Gas Technology, it would be about two and one-half times more expensive than transmitting natural gas. Still, this would be far cheaper transmission than the high-voltage power lines used today for electricity. In addition, most modern electrical transmission lines cannot move electricity more than 1000 miles without serious energy losses. These losses are one of the factors that set a limit on the size and location of modern electrical power plants. But gas transmission shows very small energy losses, even over continental distances. Therefore conversion of electrical power into hydrogen will permit the construction of very large and highly efficient electrical power plants in extremely remote areas, because the energy will be transportable over long distances with very low losses.

It must be remembered, though, that there are energy losses in converting electricity into hydrogen, and these losses could wipe out any advantage gained in the transmission system. But the *hydrogen concept* is still highly attractive, and will become more so with improved technology. Electricity and hydrogen complement each other and together offer man the best approach to an energy-oriented society, especially when one considers the many other tasks that can be performed by hydrogen.

Hydrogen plus carbon dioxide (which could be extracted from carbonate rocks) can synthesize methanol, which can be used as a fuel. With nitrogen from the air, it can produce ammonia, which is important in the production of fertilizer. Some industries that now use coal or oil as a fuel could switch to hydrogen. Blast furnaces for producing iron from iron ore could use hydrogen instead of coke by injecting a hydrogen-air mixture into a furnace charged with relatively pure iron ore. This would produce iron of a higher quality since coke contains detrimental impurities such as silicon and sulphur, and the exhaust from the plant would be a lot cleaner. In fact, this is already beginning to happen: the hydrogen process is in use at U.S. Steel, Armco and Hojalata y Lamina of Mexico. The Mexican company has reported that capital costs have dropped and productivity is higher than would be possible with conventional blast-furnace methods. The uses of hydrogen are quite varied: it is a component of chemical feedstocks, food-stuffs and petrochemicals, and it is a superclean fuel for home heating, hot-water heating and cooking, and has enormous potential in the area of transportation.

As we deplete our natural resources of petroleum, gasoline lines will begin to grow longer. And when we reach the head of the line, we will be paying a great deal more for gas. We complain about 70 cents per gallon now. It is not unrealistic to envision gasoline costing $2 per gallon in five or six years, unless we develop a substitute approach to the fuel, our mode of transportation, or both.

Luckily, the automobile does not have to run on gasoline. There are other alternatives; the electric car is one, hydrogen-fueled cars another. In 1969 Dr. Roger J. Schoeppel of Oklahoma State University demonstrated that hydrogen could be used as a fuel in a conventional internal combustion engine. He showed that, in fact, hydrogen was a superior fuel to gasoline, burning more efficiently in a lean fuel-air mixture. Hydrogen burns cooler because of its special type of nonluminous flame. It apparently radiates less heat energy to the walls of the engine cylinder. Furthermore, hydrogen allows an engine to operate at a much higher compression ratio, which results in more horsepower. Under well-controlled conditions, an engine running on hydrogen should be completely pollution-free. One of the early problems with hydrogen-fueled engines was preignition knock; that is, the hydrogen would instantly burst into combustion as it entered the engine cylinder. The result was a knocking sound due to poor ignition timing. But it was found that this problem could be greatly reduced, or even eliminated, by injecting excess air or water into the cylinders along with the hydrogen. (Incidentally, this was one of the serious problems that plagued the German hydrogen-fueled cars of the 1930s.) Dr. Schoeppel is continuing his research on hydrogen-fueled auto engines, fully convinced that a reliable efficient car engine that runs on hydrogen can be mass produced. Of course, it is not necessary to burn hydrogen in an internal combustion engine; a fuel cell could be used instead. As was discussed in Chapter 3, cars have been designed that operate on electricity, with individual motors powering the four wheels. The electricity to run the motors comes from a central power source, either batteries or fuel cells. The simplest and most efficient fuel cells operate on hydrogen and oxygen. So the electric car powered by a fuel cell fits in well with the hydrogen economy. The use of hydrogen in a car engine is really not a problem. The big holdup is the safe storage of enough hydrogen in a car to give it a practical range.

If liquid hydrogen were used, the car would require a 50-gallon tank to go as far as a conventional car can go on a standard 20-gallon tank of gasoline. It would be difficult to design a family car with room for a huge 50-gallon hydrogen tank (hydrogen requires about three times the volume of gasoline), so the tank is the big problem.

Obviously, a steel tank containing hydrogen gas under pressure would be totally impractical. A liquid hydrogen tank would be extremely costly and difficult to handle. Three possibilities would have to be considered if a cryogenic tank were used on a private automobile. First, the entire fuel tank could be replaced when it was empty. In other words, when one stopped at a gas station, instead of "fill it up," one would say "replace the tank." The empty tank would be disconnected and a full one snapped in place. Needless to say, fuel tanks would have to be standardized since a service station could not store dozens of types of large gas tanks. The exchange of tanks would also be a time-consuming process requiring skilled personnel and special equipment. Then there would be the problem of knowing how much gas is left in the tank. Certainly a driver would not want to operate his car until the tank was empty, and if there was gas left in the tank, how would one determine its contents accurately? It could be weighed—but that would take additional time and equipment. It would also be difficult for a station to determine how many tanks to keep in stock. The second possibility is to allow only highly trained cryogenic specialists to handle the refueling of cars, but it would be extremely difficult to find or train the hundreds of thousands of specialists that would be necessary. The third possibility would be to design fail-safe simplified liquid hydrogen refueling systems that could be operated by untrained (or minimally trained) service station personnel. But here again, the equipment required would be extremely expensive. Most likely, hydrogen-fueled cars will become practical only through the use of a more stable storage system such as the metal-hydride tank.

Present research and development indicate that a hydrogen-fueled car with a range of 300 to 400 miles is a definite possibility. However, because of the sophistication of the fuel system, it is most likely that hydrogen fuel will get its start in larger vehicles, replacing diesel fuel in large cross-country trucks and trains. It could also be used in long-distance buses. With this kind of experimentation, manufacturers will gain valuable experience before attempting to mass-produce inexpensive hydrogen fuel systems. Eventually all large vehicles will be able to operate on this pollution-free fuel.

In 1972 General Motors sponsored a unique intercollegiate Urban Car Design Contest, held at the General Motors Proving Ground. Sixty-three experimental cars entered the competition. They ran on a variety of energy sources including batteries, ammonia, propane and various combinations of gasoline. Two of these cars operated strictly on hydrogen; they emitted less pollutants than any of the other cars. One was a Gremlin converted to run on hydrogen by a team from the University of California. It came in a close second to a converted Volkswagen, whose exhaust was so clean that if it had been driven in the city, the exhaust would have been purer than the air it was drawing in. The automotive industry is certainly a major area for the application of hydrogen as a fuel source.

Another area that is rapidly becoming a prime candidate for hydrogen fuel is aviation. Although jet fuel presently represents only a small part of U.S. petroleum consumption, the aviation industry will eventually grow to the point where it will demand as much fuel as the automobile industry. Furthermore, jet fuel, which is highly refined, costs much more than crude forms of fuel oil. So because of fuel costs and limited supply, hydrogen will eventually become very competitive in the aviation industry.

Its light weight (hydrogen is the lightest of all the elements) and high energy content make hydrogen extremely attractive to the airlines. These advantages hold a special interest for designers of high-speed and long-range aircraft, where fuel weight is a major problem. Engineers at the Lockheed Aircraft Corporation estimate that the takeoff weight of an advanced supersonic transport could be reduced by about 200,000 pounds if hydrogen were used as a fuel. Converting present-day jumbo jets to bulky hydrogen fuel means finding extra space for additional fuel volume. This could be accomplished by the addition of wing tanks. But two factors must be considered. One, additional insulation would be required to thermally protect supercold liquid hydrogen, and this would add a weight penalty. However, a hydrogen-burning jet engine runs more efficiently, giving more thrust per volume of fuel, and this would more than offset the extra weight. Second, if hydrogen were used, large high-speed aircraft could be built of aluminum instead of heat-resistant steel alloys since the liquid hydrogen could be used as a coolant, thereby performing two operations: the hydrogen would be preheated before injection into the engine and the aircraft would be protected from frictional heat at supersonic speeds. And the aluminum structure would represent a further weight reduction. The structural weight and fuel weight savings would result in a commercial aircraft with a range two and a half times that of the same type fueled with present jet fuel.

Jet engines are easily converted to hydrogen. In fact, because hydrogen burns cleanly and because the supercold fuel acts as a coolant, the engine can be run at a higher, more efficient temperature. Furthermore, engine parts would last longer running on hydrogen, which would mean lower maintenance costs. And there is still another advantage—noise reduction. For supersonic transports, a lighter, higher-flying aircraft will produce lower-intensity sonic booms, and because of design advantages, a plane will be able to take off with its engines partially throttled. So hydrogen-fueled planes will mean a lot less noise around airports. There are obstacles, such as bulky fuel areas and fueling problems connected with liquid hydrogen at −423° F., but these can be overcome with a reasonable amount of research and development. Hydrogen will make airliners more effective, efficient and pollution-free. This fuel can even allow us to operate supersonic aircraft at high altitudes without damaging the protective ozone layer of the earth. Eventually, we will have no choice in the matter. In 1970 aviation accounted for approximately 12 percent of the fuel consumed by transportation in the United States; this figure will rise to 27 percent by 1985, and by 2000 the aviation industry will require about 32 percent of all transportation fuel. Meanwhile, all U.S. production of petroleum will continue to drop. The handwriting is on the wall—let us hope the right people can read it. Once aircraft and other transportation systems change over to hydrogen, there will be a 25 percent reduction in fossil fuel consumption. And the transportation industry will be set for the next thousand years or so.

We are fast running short in other areas of energy. U.S. petroleum production continues to drop, and imports are steadily increasing. The OPEC nations now control about 35 percent of our oil supply, and our reliance will continue to grow. The combination of government controls to reduce oil consumption and the domestic shortage will lead to much higher fuel prices. Natural gas, which heats about 55 percent of the nation's homes, is used for petrochemical feed stock and fertilizer and is the largest source of energy for industry, is even scarcer. A lot of jobs have already been lost because of the gas shortage. The shortage forced industry to use an additional 400 million barrels of oil between August 1974 and August 1975, further increasing our imports. At the present rate of depletion, most of our accessible natural gas may be used up by 1980, a situation that could have disastrous effects on our economy. As our supplies of natural gas and oil run out, and as their cost climbs, hydrogen will become economical and very attractive. Once the production methods are improved and hydrogen is produced in large quantities, its price will steadily fall until we have cheap, clean energy in abundance. This abundance will be manifested in a huge amount of stored energy representing billions of revolutions of modern windmill blades, hundreds of rising and ebbing ocean tides and dozens of days of bright sunshine. These renewable forms of energy will become practical once hydrogen is an effective storage medium. And hydrogen will be directly translatable to electricity. Together with liquid oxygen (which will also be available in abundance from electrolysis), it will be used as a clean fuel for MHD (magnetohydrodynamics) and fuel cell electrical generating plants. Hydrogen could even be used as a fuel in present-day gas turbine generators. The conversion of these turbines to hydrogen would not be difficult, and we would have cleaner and more efficient operating power plants.

There are those who will recall the huge zeppelin named the *Hindenburg*. In 1937 the airship was carrying happy tourists from Europe. Just as it was preparing its final descent to the docking point in Lakehurst, New Jersey, it burst into a blinding, all-consuming fireball. Ever since that grisly accident hydrogen has been perceived as a death gas (the Hindenburg Syndrome). The fact is, all fuels can be highly dangerous. Hydrogen, when mixed with oxygen in a confined situation, can form an explosive mixture. It has a low ignition point and creates an intensely hot, colorless flame that burns very rapidly. But in the open air or in well-ventilated areas, leaks or spills of liquid hydrogen defuse and disperse rapidly (because hydrogen is so light). The risk of ignition and fire is less than for gasoline, and hydrogen explosions are actually quite rare. It should be recalled that for about 100 years most people safely lit their homes with town gas, which contained up to 50 percent hydrogen. Of course, hydrogen requires adequate ventilation, leak prevention and the elimination of open flames—but so does natural gas. Gasoline fumes in a gas tank can explode with bomblike destruction. Yet we manage to relax while carrying 20 gallons of this highly flammable liquid at 70 miles per hour in our automobiles. Gasoline creates fumes that mix with the air in our auto tanks. One spark in the tank and—*wham!*—a bomb goes off. The hydrogen in a pressurized tank cannot burn or explode, so it is safer than gasoline. According to D. B. Chelton of the Cryogenic Engineering Laboratory, hydrogen ranks even ahead of common kerosene on most safety counts.

Still, most new things evoke fear. When automobiles began operating in England, a special ordinance was passed that required the vehicle to be preceded by a man on foot carrying a red flag. Similar fear laws were passed in the United States. When the first tank trucks began transporting hydrogen on U.S. roads in 1956, they had to be escorted both front and rear by jeeps painted red—a manifestation of the Hindenburg Syndrome. Today huge amounts of liquid hydrogen are moved across the country on a routine basis by tank trucks and railroad cars. Also, NASA has been handling liquid hydrogen for nearly 20 years without a single major accident. Other large companies such as Union Carbide and Foster Wheeler have been operating huge hydrogen production plants and storage facilities for years with perfect safety records, and these plants are producing hundreds of millions of cubic feet of hydrogen each day. Although the general fear of hydrogen started with the *Hindenburg* disaster, few people know that as spectacular as the incident appeared, the fire was almost over within two minutes and of the 97 people on board, 62 survived. Contrast that with a modern airliner crash; the gasoline can burn for hours, leaving virtually no escape for passengers. Herein lies one of the great advantages of hydrogen: with proper ventilation, liquid hydrogen will boil off rapidly. Furthermore, once it ignites, it expands and rises with great speed, leaving the vehicle relatively cool, as opposed to gasoline which, because of its weight and molecular structure, clings to the ground and the burning vehicle and consumes all the material within this area. So a passenger in a hydrogen-fueled aircraft might very well stand a better chance of surviving a fire.

Perhaps the biggest disadvantage of hydrogen is that only a small amount of energy will cause it to ignite. It takes only one-tenth the energy needed to ignite gaseous gasoline (mixed with air)—which is to say a static electric spark can do it. But with a sealed positive-pressure tank, which contains no air, danger from static sparks can be eliminated. On the other hand, in his published research work, *Dangerous Properties of Industrial Materials,* N. Irving Sax shows that hydrogen must be raised to more than twice the temperature of gasoline before it will ignite (1085° F. for hydrogen versus 495° F. for gasoline). In other words, under the same temperature conditions, it is more difficult to ignite hydrogen than gasoline. Sax also points out that since gasoline is slightly poisonous, it produces slightly toxic fumes during a fire. Hydrogen is completely nonpoisonous.

Hydrogen is extremely difficult to detect when it leaks because it has no color, odor or taste. But this danger can be overcome by adding an odorant to it (as is done with natural gas) and a material that adds luminescence to its flame. Surprisingly, the most common danger with liquid hydrogen is frostbite. But this occurs only when large amounts of liquid hydrogen come into contact with the body. Fortunately, a small amount of liquid hydrogen falling on human skin acts like droplets of water on a red-hot surface; it sputters and immediately flings itself from the skin, before it can do any damage. Hydrogen-fueled vehicles could be designed to dump their fuel in the event of an accident. This could never be done with gasoline, since it would not evaporate with the rapidity of hydrogen. To sum up, fire is the greatest danger with gasoline; with hydrogen, it is frostbite.

One of the delightful aspects of the hydrogen economy would be unconcern for fuel spills. Today oil spills are a serious threat to our environment. They are deteriorating our bodies of water and ruining our beaches. And large oil spills are nearly impossible to clean up. Their effects last for years or even decades. In the event of a large hydrogen spill, the fuel would quickly evaporate, leaving no trace behind.

Hydrogen and electricity—a perfect combination that works in complete harmony with nature. Hydrogen will complement the electric society, giving it flexibility, permanence and consistency. There are no serious technical or economic impasses. When we finally reach the point of massive hydrogen utilization, this world will be a much cleaner place. It will also reap the great benefits of nature's most abundant and efficient fuel. We look upon hydrogen as a wonder of the future society; but there have been those in the past who have had the magnificent foresight to predict this series of events. In 1874, through the voice of one of his forward-looking characters in *The Mysterious Island*, Jules Verne leaped more than a century into the future: "I believe that water will one day be employed as a fuel, that hydrogen and oxygen that constitute it, used singly or together, will furnish an inexhaustible source of heat and light." Now we are ready for that movement. Compelled by nature and animated by wisdom and necessity, man must make this vital transition. A by-product of the systems described in the next five chapters will be hydrogen in sufficient quantities to maintain our viability as long as we can survive on this planet as social beings.

6

Oceans of Electricity

With a consistency that has defied man's most ingenious analysis, the oceans of the world advance and recede from the land with amazing regularity. It is this seemingly perfect consistency that has puzzled and fascinated man from the first moment his glance fell upon the open sea. His many attempts to unravel this mystery have not been very fruitful; in the words of Dr. Irving Michelson, "The daily rise and fall of the ocean surface, the very heartbeat of the oceans themselves, is a subject of which our fundamental knowledge ranks somewhere between meager and nil." But whether or not we ever fully understand this force of nature, it represents a formidable source of energy, so the regularity and reliability of the tides intrigues many scientists and engineers searching for new ways to harness clean, renewable energy.

There have been people in the past who did not require extensive scientific analysis, engineering developments or economic studies to justify the use of tidal power. In their ignorance they went ahead and built devices to make the tides work for them, and they worked quite well. As long ago as 1100 A.D. tide mills or paddlewheels driven by tides were used to grind corn in Britain and France. Later, in colonial America, farmers trapped seawater in small estuaries at high tide by means of a small dam; when the tide receded, the water escaped through a waterwheel and generated about 50 hp of energy. Unfortunately, because of the availability of other forms of energy, tidal power never made the transition from a small farmer's helper to a huge electrical energy generating plant. But men of science have remained interested in the cause of the tides' consistency. This challenging mystery led the great mathematician P. S. Laplace to lament, "Tides represent the most intricate and complex problem in all of celestial mechanics."

This mystification is not shared by the animal world. The California grunion seems to have a perfect knowledge of tides. These little creatures synchronize their spawning time with the tides, coming ashore just after the peak tides have begun to ebb. There is only a brief period of time when the eggs can be laid in the sand where the tide will not disturb them before they hatch ten days later. The male and female grunion are unerringly punctual; their knowledge of the rhythm of the tides leaves nothing to chance.

The word *tide* refers to the periodic and persistent rise and fall of the sea level twice each day, with an average of 12.25 hours lapsing between consecutive high tides. Therefore the interval 24 hours, 50 minutes, considered the average period of time between two successive risings of the moon, is the key to the definition of tides. A tide can be defined only in terms of the natural forces which caused it, and these causes are astronomical in character. For example, the expression *tidal wave* is really a misnomer. A tidal wave, that is, a wave of unusually great size, arises from earthquakes or from extreme weather disturbances. Therefore it is not tidal in nature but a totally different natural phenomenon. Although we do not notice them, tidal forces exist in midocean; they manifest themselves as the rising and falling of the ocean waters. In midocean the tidal energy (force)

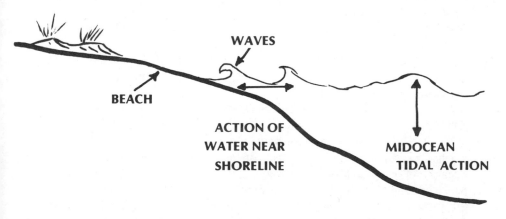

WAVES

BEACH

ACTION OF WATER NEAR SHORELINE

MIDOCEAN TIDAL ACTION

FIGURE 36. Tidal forces occur in midocean as water rises and falls, and at the seashore as inrushing and outflowing water.

is transferred sideways as a series of shifting swells. No large amount of water is transferred from one point to another (see Figure 36); the water simply rises and lowers. But near the land areas, at the beaches, as the water piles up, it topples over in the form of waves and rushes onto the beach. As long as the surface of the ocean continues to rise, water rushes onto the beach, raising the level of the waterline. We refer to this action as the incoming tide. Its highest point is called high tide, and naturally, the condition of the ocean reaching its lowest level is referred to low tide. The difference in the level of the water between low tide and high tide is called the tidal range.

WHAT CAUSES TIDES

Many forces affect the waters of the earth—the rotation of the earth; wind, which makes water choppy; barometric pressure variations, which exert different degrees of pressure force upon the surface of the water; and earth tremors or earthquakes, which can disturb large amounts of water. Most of these forces are not cyclical; they do not occur in a consistent pattern. The tides, on the other hand, are extremely consistent throughout the year. This is because the forces that cause tides are themselves very consistent. These forces are the gravitational attraction of the moon and the sun upon the oceans. Actually, these gravitational forces act upon all bodies of water of the earth, but their effects upon the oceans are very noticeable because of the huge sizes of these bodies of water. About three-quarters of the earth is covered by water; it is evident, therefore, that the forces necessary to move such massive bodies of water must be very powerful. Once set in motion, the oceans of the world contain vast amounts of potential energy. In terms of electricity this potential energy exists in millions of megawatts—certainly enough to contribute significantly to the elimination of man's energy crisis. How much of this energy is available for man's use? What methods can extract it? Is it practical and economical to harness energy from the oceans?

Many scientists who have been studying these questions have turned their attention to the most familiar of the ocean forces, the tides. The moon, which is our closest celestial neighbor (about 240,000 miles from earth), exerts the most powerful of all gravitational forces upon the oceans. As it moves around our planet, its attractive force literally pulls the oceans of the earth slightly away from the planet. In other words, it raises the surface of the oceans, causing a slight bulge in the water level on the side of the earth facing the moon as well as on the opposite side (as seen in Figure 37). It is this raising of the water level that results in the tides. Since the water rises at two points on the earth (A and B in the figure),

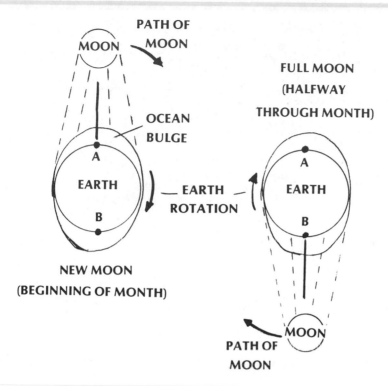

SUN   SUN

PATH OF
MOON   MOON

MOON

FULL MOON
(HALFWAY
THROUGH MONTH)

OCEAN
BULGE

A   A

EARTH   EARTH   EARTH
ROTATION

B   B

NEW MOON
(BEGINNING OF MONTH)

MOON

PATH OF
MOON

FIGURE 37. Gravitational effects of the moon on the earth's oceans.

two areas in the world experience high tides simultaneously. All points in between experience varying degrees of tidal effects. But the earth is rotating, so that point A will have a high tide when it reaches the opposite side (B) about 12 hours later. Of course, the moon is also traveling in a path around the earth. Therefore, because of its change in position relative to the sun, its influence will vary at different times of the month. Variations in tidal effects are also due to the fact that the distance between the moon and the earth changes throughout the month. One may ask why the waters bulge on both sides of the earth since gravitation is an attractive force. The answer is rather complex and involves a subject called vector analysis. But to put it simply, a combination of forces from the gravity of the earth, the moon and the sun cause the effect illustrated in Figure 38. These forces tend to push down on the sides of the earth, thereby lowering the level of the oceans. A similar effect can be observed by taking a rubber ball and squeezing it on two sides. It will be noticed that the ends of the ball (perpendicular to the points of pressure) will be extended slightly. This is

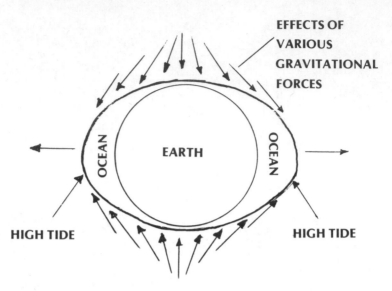

FIGURE 38. A complex combination of gravitational forces from the earth, moon and sun cause two high-tide points on the earth.

basically what occurs as a result of the various gravitational forces acting on our planet.

The sun also influences the tides. Although much farther away (93 million miles) than the moon (240,000 miles), its great size (about 900,000 miles in diameter) generates a powerful gravitational force that has almost the same effect on the oceans as does the moon's. The sun's force adds to the complexity of the situation, and its influence varies as the earth follows its great ecliptic route about the sun. Because of the many motions involved (the earth's rotation, the moon's path about the earth, and the earth's route around the sun), the size, duration and interval spacing of the tides are constantly changing. As seen in Figure 39, there are three key configurations that either maximize or minimize the tides.

The first is when the moon is positioned on the opposite side of the earth—this is known as the full moon. In this position the moon, earth and sun are aligned as a single axis (in a straight line); therefore the gravitational forces have a maximum effect on the oceans and result in very high tides called spring tides. The second configuration is when the moon is positioned directly between the earth and the sun (new moon). This position also results in maximum tidal forces and spring tides. The third configuration of these celestial bodies occurs when the moon is perpendicular (at a 90° angle) to the earth-sun axis. In this position the gravitational forces of the sun and the earth partially cancel each other. Therefore the tides that result, called neap tides, are only one-third as high as the spring

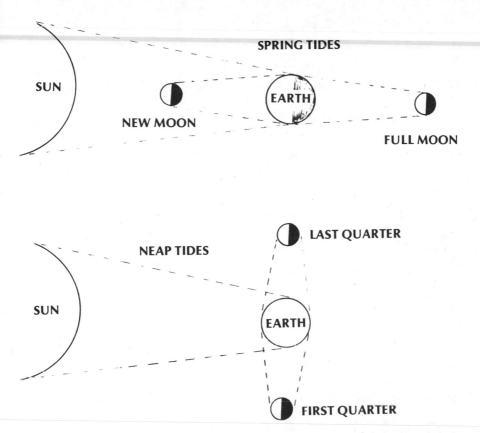

FIGURE 39. Relationship of the combined gravitational forces of the sun and the moon upon the earth's oceans.

tides. At all other times of the month the size of the tides reflects the varying positions of the sun and the moon in relation to the earth.

Although most areas of the earth experience two tides each day, and although the average time between succeeding tides is about 12 hours and 25 minutes, the tidal variations around the earth are quite significant. In some areas of the East Indian Seas two successive high tides are separated by about 12 hours while in the China Sea they may be separated by more than 24 hours. In other places, such as Southhampton, England, high waters (high tide) are often doubled—that is, the tide reaches its peak, begins to recede for a short period of time (dropping its level just slightly), and then a second tidal rush enters, bringing the tide to its maximum again. At other places the opposite occurs and the low tide is doubled.

The tides of the world vary considerably in range. World averages for spring tides are about 13 feet 9 inches and for neap tides about 7 feet 9 inches. However, these are only average figures; great departures exist in strong tidal areas. At the Bay of Fundy between the Canadian provinces of New Brunswick and Nova Scotia, which is the site of the world's greatest tidal range, the difference between the water level at low and high tides is 50 feet; in the Mediterranean Sea the tides are barely detectable. The magnitude of tides can sometimes be reinforced by the wind; in extreme cases strong winds can modify the size of a tidal rise by as much as five feet. Even changes in barometric pressure can affect the size of tides. The forces involved in tidal formation are highly complex in nature, but one fact remains clear: there is a huge amount of energy in the rushing tides that can be harnessed by man. If we could extract this energy economically, we would have a source of energy that would last forever and be perfectly in tune with nature. Tidal power generates absolutely no pollution, requires no fuel, is available with no political complications and is perfectly safe. But there are two major limiting factors regarding tidal power.

One of these is cost of construction, which can run into the billions of dollars for a large project. The other is available sites where tidal ranges are great. As far as construction costs are concerned, the real cost of any power-generating station must include a number of associated items such as fuel, manpower requirements and the effects of the power plant on its surroundings, such as generation of additional traffic for fuel deliveries, pollution and security requirements. After all factors are considered, including eventual increases in fuel prices, the capital cost of a tidal power electrical generating station is not very much out of line. And in the long run it will be one of the finest forms of power generation we could incorporate into our society. As stated by J. Hilbert Anderson, president of Sea Solar Power, Inc., "It's inevitable. There's no way we can avoid using the energy from the ocean. The ocean will one day be a major source of energy."

Although tidal power sites don't exist in great abundance, it is still worthwhile to use what sites do exist in the United States and around the world. Besides the Passamaquoddy Bay area (Bay of Fundy), there are such sites as the tidal estuary of the River Severn in England, Mont St. Michel in northern France, the Bay of L'Aver Vrack in Brittany and the Gulf of San José in Argentina. The Yellow Sea off the west coast of the Republic of Korea, near Inchon, has a tidal range of 13 to 26 feet. Since this country is devoid of petroleum, natural gas and coal, tidal power could provide a valuable source of energy. This is also true in another remote part of the world, northwestern Australia at a point called Collier Bay. Here, in a narrow pass through which Walcott Inlet empties into Collier Bay, run some of the fastest currents in the world, with a tidal range that regularly reaches 36 to 40 feet. It is an excellent site for the production of pollution-free electricity. Two other tidal sites must be mentioned because they are presently harnessed and supplying a significant amount of clean energy. One electrical generating plant is located in the U.S.S.R. (completed in 1969) in the Kislaya inlet on the shore of the White Sea. The other is on the Rance River estuary in the Gulf of St. Malo, on the coast of Brittany. These are the major tidal sites that have been categorized as most likely points for power generation. But there may be many others, both in the United States and around the world. Even though they are not as large, the fact that tidal power is totally nonpolluting may make them a valuable source of energy. The question is, how does one go about harnessing the tides?

THE WAY IT WORKS

The whole concept of tidal power is based on the fact that a great deal of water rushes from the ocean to an inland area, which may be either a bay or an estuary (see Figure 40). At one time of the day the water level in an estuary may be only 10 feet or so deep. As the tide comes in, the estuary becomes flooded and the water level begins to rise. Under extreme cases it may change from 10 to 60 feet. Shortly after the level has reached its maximum, the tide begins to recede and after a number of hours it returns to its low point again. During the course of these actions billions of gallons of water will have flowed into and out of the estuary (or bay). This huge amount of moving water can perform a lot of work, such as rotating large turbine wheels. This is the key to tidal power.

A tidal power generating station consists of a huge damlike structure in which are located large sluice gates (doors to let in water). In one tidal power method (shown in Figure 41A), water is allowed to flow into the estuary through the sluice gates. The incoming tide raises the water level on the estuary side of the dam. Once the tide has reached its peak, the sluice gates are closed, trapping the water behind the dam. On the ocean side of the dam, the water level falls as the tide goes out. When this level is low enough, special water control gates are opened, allowing water to rush into a duct that leads to one side of a large turbine wheel. The great power of the rushing water spins the turbine wheel. A drive shaft connects the turbine to an electrical generator, which produces electricity. Obviously, the system will continue to generate power for as long as it takes the trapped water to pass through the water ducts in the tidal power dam, which would be six hours on the average. The disadvantage of such a system is that it generates electricity only during one-half of the tidal cycle.

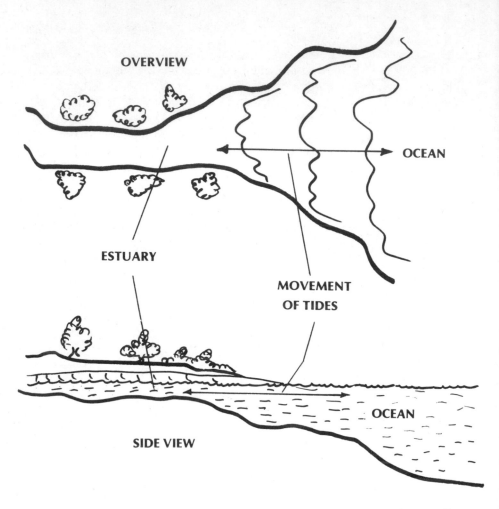

OVERVIEW

OCEAN

ESTUARY

MOVEMENT
OF TIDES

OCEAN

SIDE VIEW

FIGURE 40. Tidal power is based on the fact that large volumes of water flow into and out of bays and estuaries.

By using a reversible turbine assembly (one driven by water traveling in either direction), both halves of the tidal cycle can be put to use. This method would work as follows (see Figure 41B and C). As the tide begins to rise, instead of sluice gates allowing the water into the estuary, the water is allowed to flow through a water inlet control gate that channels the incoming tidewater to the turbine wheel. This operation continues while the tide is rising and remains above a specific level. Once the tide has fallen below this point, the inlet control gate is closed. When the tide has fallen to a low enough point, a water-control gate on the opposite side of the tidal dam is opened and the trapped water is allowed to flow through the turbine

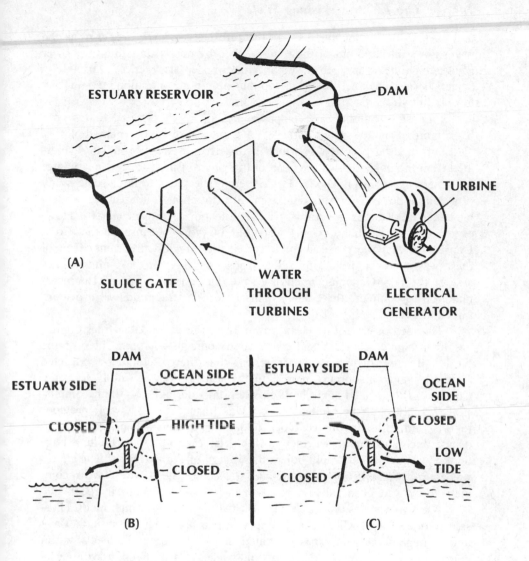

ESTUARY RESERVOIR

DAM

TURBINE

(A)

SLUICE GATE

WATER
THROUGH
TURBINES

ELECTRICAL
GENERATOR

DAM

ESTUARY SIDE

OCEAN SIDE

CLOSED

HIGH TIDE

CLOSED

(B)

ESTUARY SIDE

DAM

OCEAN
SIDE

CLOSED

LOW
TIDE

CLOSED

(C)

FIGURE 41. Electrical power can be extracted from complete tidal cycles.

and back to the ocean. In this way the turbines can extract energy as the water flows into and out of the estuary, thereby nearly doubling the power generation time. Can such a system work in actual practice? Absolutely. Scientists and engineers have no doubts about its operation and in fact have built two such systems, as mentioned.

Interest in building a tidal power system in the United States goes back more than fifty years. In 1919 a Canadian engineer named W. R. Turnbull made a proposal in which hydroelectric power would be generated from the huge tides at the head of the Bay of Fundy. A few years later an American engineer, Dexter P. Cooper, made the first large-scale study of potential power from this tidal area. But not enough interest was shown by either industry or government to get the power project started. However, by 1935 so many people were out of work because of the Great Depression that the government began instituting job programs. One of these was a power project to utilize Cobscook Bay on the United States side of the U.S.-Canadian boundary line. The Army Corps of Engineers managed to complete three small dams before federal funds were discontinued in 1937.

The idea of extracting clean power from the dependable tides continues to fascinate scientists and government officials on both sides of the border. Because of a growing awareness that energy should be extracted from all available sources, interest in the tidal power of Fundy Bay and Cobscook Bay peaked again in 1948 when an International Joint Commission (Canada and the United States) was formed to review all previous reports and estimate the cost of conducting a comprehensive study to decide once and for all whether it would be economical to build a huge tidal power plant that would utilize the energy of the two bays. The report was completed in 1950 and work was started on the survey. Two recommendations were eventually issued—one in 1956 and a second in 1959.

The reports describe a system of two "pools"—a high pool (Passamaquoddy Bay), which would empty into a low pool (Cobscook Bay) and 90 huge filling gates. It is estimated that 70 billion cubic feet of water enter and leave these bays during a tidal cycle. This tremendous volume of water would be channeled through 100 turbogenerators. About seven miles of rock-filled dams would be needed to form the two pools. It was estimated that such a system could generate about 1150 MW of electrical power. The construction challenges would be massive. The seven-mile dam would have to be built in water ranging from 125 to 300 feet in depth and moving at a rate of up to 20 feet per second. About 459 million cubic feet of clay would have to be moved and dumped at the dam site to form its core, which would then be covered by millions of tons of stone. The massive housings for the 100 turbogenerators would each measure 55 feet wide by 188 feet long and be built of reinforced concrete. The heavy steel flood gates, which would open upward, would measure 30 feet by 30 feet. To protect the turbines from the corrosive effect of seawater, all metal parts would be made of corrosion-resistant alloys, protective coatings would be used and a special system called cathodic protection would be incorporated to prevent decay (through electrolysis) of metal parts. This ultramodern power plant would practically run by itself. The synchronization of electrical loading (the amount of electricity drawn from a generator), and the starting and stopping of each electrical unit, would be automatically controlled by computers. The computers would base their decisions upon tidal predictions.

In 1956 the IJC (International Joint Commission) reported that such a power project was technically feasible. On the basis of this finding, a five-year investigation of the economics of the project was initiated (the 1959 report was a preliminary evaluation). On April 4, 1961, the final report was released and the verdict was negative: "It is evident that construction of the tidal power project by itself is economically unfeasible by a wide margin. . . .

In short, the Commission finds that the tidal project, either alone or in combination with auxiliary sources, would not permit power to be produced at a price which is competitive with the price of power from alternative sources." The IJC recommended that development of the project might be possible when less costly energy sources in the area were exhausted. But it did not end there.

On May 20, 1961, President Kennedy requested that the Department of the Interior form a new study group to review and evaluate the IJC report and advise him of any changes in fuel, engineering and financing costs that might make the tidal power project economically feasible. In December 1961 the new group reported that such a project, if used as a source of peak power, would be economically feasible. In the Public Works Bill for fiscal 1963 Congress voted a sum of $200,000 for another study of the feasibility of building a peaking power plant. One of the Interior Department's recommendations was that the President instruct the Secretary of State to initiate negotiations immediately with the government of Canada for a sharing of power that would assure maximum benefits to both nations.

That was over twelve years ago. Nothing of any significance has been done since, except for more economic and scientific studies. Unfortunately, there were two obstacles to undertaking this tidal power project. One was the absence of electrical power storage technology. The Bay of Fundy project obviously had to be synchronized with the tides. Since the time of tidal activity is constantly changing, the power plant would frequently be faced with the problem of generating a great deal of electrical power at times of the day when demand is very low. What could be done with all the electricity? It could not be stored, so it would be wasted. That obstacle no longer exists; there are now a number of excellent storage methods.

The second obstacle was partly factual and partly imaginary. It is true (this is the factual part) that other sources of inexpensive fuel were readily available, but it was shortsighted of the IJC to make a long-range forecast about the cost of generating systems on the basis of the continued availability of cheap fuel. Today's fuel costs are invalidating a lot of system comparison studies. Now comes the imaginary part of the obstacle. The commission failed to consider other economic variables besides cost of fuel—namely, the hidden costs of pollution resulting from fossil-fuel-based power plants. Erosion of our structures, illness, destruction of land and water, deaths incurred in extracting fossil fuels from the earth—all these factors have to be taken into account when evaluating a power generating system. Had they been seriously considered by the IJC, the Bay of Fundy might be in operation today. As far as the other problems are concerned, such as high-speed currents, water depth, effects of seawater, they are solvable, even simple, compared to the mind-boggling problems facing the designers of atomic breeder reactors. With modern technologies, it is possible that as much as 4000 MW of electrical power could be made available to the southeast portion of Canada and to Maine, New Hampshire, Massachusetts and possibly Vermont. The growing awareness of our dependence on the ecological balance of nature and the recent recognition of the tenuousness of our fuel supplies are making projects such as tidal power look more attractive and causing renewed interest in the Bay of Fundy project.

While we were busy with countless studies and evaluations, on the other side of the Atlantic, in Brittany, the dream of harnessing tidal power has come true. In June 1956, some 36 years after the United States first considered a tidal power project, the Société Hydrotechnique de France held a general conference in Paris. Among the technical papers presented was the first detailed proposal for harnessing the tidal flow of water in and out of the Rance River estuary on the northern coast of France. The idea appealed to the French government, and eleven years later electrical power was being generated from a plant that operated on the energy of the daily tidal flow at the Rance estuary.

Although the Rance project is nowhere near the size of the project considered for the Bay of Fundy, it was nevertheless a massive undertaking that cost the French government (back in 1967) $80 million to design and construct. The Rance power system is made up of 24 turbines, each of which produces about 10 MW of electricity. The site is ideal for tidal power because it is a narrow estuary with a tidal range of approximately 44 feet. The Rance station has been designed for peak power conditions and takes advantage of the energy in the complete tidal cycle, using both incoming and outgoing tide. La Rance (as it is called), which is 2300 feet wide and 85 feet high, is among the twelve most powerful hydroelectric power stations in France, having a total power output of 240 MW.

One may wonder why fewer turbines of larger size were not used in place of the 24 small turbines. There were certain advantages to this approach. First of all, the effective hydrostatic head (water pressure) is greater on a smaller turbine assembly and therefore results in greater efficiency. Also, with a larger number of smaller turbines, the breakdown of one unit does not seriously affect total electrical output. Furthermore, small modular units can be removed and replaced in a matter of hours and spare units are economically practical. Finally, when a large number of identical generating units are produced, they can be manufactured using economical assembly-line methods. All these advantages can be applied to the Bay of Fundy project.

The La Rance power project can be considered a pilot program, a test site. It was conceived and executed on a small, controllable and economical scale, but it proved tidal power stations can work and it provided valuable experience and know-how that can benefit the designers and builders of larger tidal projects such as the Bay of Fundy. Since tidal power represents such a clean, natural source of energy, a great effort should be made to locate other possible sites. They may be found in the northeast, the northwest or along the Alaskan coastline.

Tidal power is not the only method for getting power from the movements of the ocean. Not only does the ocean move in terms of tidal forces, but there are also a large number of complex currents, where portions of the ocean move in a continuous stream covering distances of up to thousands of miles. These currents move at different speeds and levels. At the surface a current flow may be from south to north at a few knots per hour, while a couple of hundred feet down the water may be flowing in a totally different direction at three or four knots per hour.

The force behind these currents could be put to work to generate electricity. According to three highly experienced ocean scientists (Harris B. Stewart, Jr., and John R. Apel of the National Oceanic and Atmospheric Administration, and William S. Von Arx of Woods Hole Oceanographic Institute), a good deal of energy could be extracted from the northward-flowing Gulf Stream. Their idea is to construct about 200 large turbines, which might look like underwater windmills, in the undersea area between Florida and Bimini at depths between 100 and 400 feet (see Figure 42). The speed of the current at this location is as high as 5.5 miles

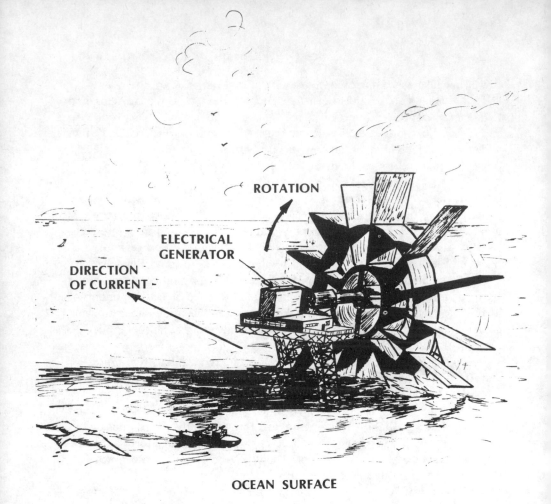

ROTATION

ELECTRICAL
GENERATOR

DIRECTION
OF CURRENT

OCEAN SURFACE

FIGURE 42. Ocean currents can be put to work generating electrical power.

per hour. If the turbine blades were positioned at the right level, the fast-moving current would rotate the huge windmill-like blades of the turbine, which would, in turn, rotate an electrical generator. The scientists admit that the idea has not been tested and that a feasibility study would be required, but it is not beyond our engineering capabilities. And there is a considerable amount of power available from this source. Stewart, Apel and Arx have calculated that if only 4 percent of the energy were tapped at that single point, as much as 2000 MG of electrical power could be generated.

There is another way of using the motions of the sea: electricity can be extracted from ocean waves. This is not as crazy as it sounds. In fact, it is a rather ingenious idea. Large bodies of water almost always experience

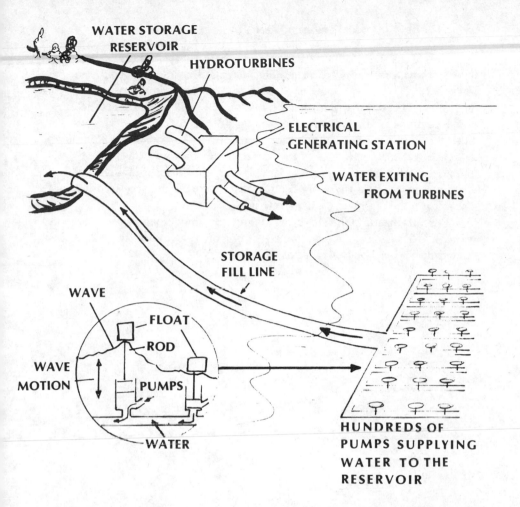

WATER STORAGE
RESERVOIR

HYDROTURBINES

ELECTRICAL
GENERATING STATION

WATER EXITING
FROM TURBINES

STORAGE
FILL LINE

WAVE

FLOAT

ROD

WAVE
MOTION

PUMPS

WATER

HUNDREDS OF
PUMPS SUPPLYING
WATER TO THE
RESERVOIR

FIGURE 43. The wave motion of the sea contains enough energy to generate electricity.

a certain amount of surface disturbance from the wind and the tides. Waves have a vertical motion—they continuously move up and down (except at the shoreline, where they break horizontally). This is what gave Charles M. Johnson the idea. Suppose one were to place a float on the surface of the water; it would move up and down with the movement of the waves. Now suppose one were to connect a rod from the float to a piston inside a stationary cylinder below the water. As the float moved up and down with the waves, it would move the piston below the water (see Figure 43). In other words, the float riding the waves could be used to operate a pump. As Johnson envisions it, large numbers of these pumps could be used to lift water up to a reservoir. The water in the reservoir could be sent down a large duct to rotate hydroturbines, which would drive electrical generators. The water would then be discharged back to the sea. The cycle would continue, with the waves supplying all the energy and with no disruption of the earth or any pollution. Except for the consumed electricity, no heat is generated by this or any other ocean motion power system. Of course, the main advantage of the last two methods discussed—current and wave motion—is that these are pretty constant forces and therefore do not generate too much power at the wrong time of day.

The important point to bear in mind is that clean, safe energy is available from the sea whether we use it or not. Enormous energy is being expended; it would be foolish of us to ignore it.

7

Reaping the Wind

On the breezy crest of Grandpa's Knob, a 2000-foot hill overlooking the Champlain Valley 12 miles west of Rutland, Vermont, a huge structure towers 12 stories into the air. Its mighty outstretched arms measure 175 feet across and it is visible from 25 miles away. This, the biggest windmill in the world, is quietly and smoothly capturing energy from the wind and converting it into electricity.

Built in 1941, this structure is the famous Smith-Putman Wind-Turbine. The massive power plant was the dream of Palmer Cosslett Putnam and the S. Morgan Smith Company, and the culmination of years of research and development. Putnam envisioned the eventual construction of a series of giant windmills that would supply large amounts of clean, inexpensive electrical energy to the citizens of Vermont.

It was a dream that rested on a good foundation; after all, the winds and breezes were following their complex courses around the planet long before the first life forms ventured out from the sea. The wind is ever present, though undetectable to the eye.

> Who has seen the wind
> Neither you nor I
> But when the trees bow down their heads
> The wind is passing by.

Christina Rossetti's graceful words echo the delightful presence of a soft breeze, but the winds of the earth can also be terrible. With violence, they can turn a placid sea into a boiling fury of angry waves, dash the sturdiest ships against treacherous rocks and tear a proud tree from the life-giving earth. In the form of twisting winds or devastating gales, by the name of hurricane, typhoon, or cyclone, these mighty masses of rushing air can leave behind a path of destruction second to no other force on earth, save the cataclysmic erruption of the earth's core.

Winds are pure, raw power, whipping around the earth in a complex array of currents, gusts and eddys that swirl along the ground, whip up the sides of mountains, rise to dizzying heights and race along at 400 miles per hour at the upper reaches of the troposphere. Until recently, their tremendous power had been all but ignored by modern man. Winds are constantly watched by meteorologists, but only to give warning of their destructive intent. The wind has been, for the most part, the bad guy, the force that hurts and kills people, destroys crops and causes untold property damage. Many people ignore the concept of harnessing the power of the wind and ridicule those ecological pioneers who would accept nature's invitation to reap the wind.

There are two common criticisms of wind power. One is that there is not enough potential power in the wind to make a meaningful contribution to the energy needs of mankind. Wind power, assert its critics, could never produce enough electricity to make development and construction projects worthwhile. The other criticism is that wind is totally undependable as a source of energy. There is no way of knowing exactly when the wind will suddenly die down; it blows in fits, gusting one moment and barely present the next. This is certainly not a method of power generation that a community would want to rely on for its basic energy needs. Both criticisms sound perfectly logical. However, the facts of the first are incorrect, and the conditions leading to the second are alterable with modern technology. Wind inconsistency and the modern approach to this problem will be discussed later in this chapter. Let us now examine the first criticism.

No part of the earth's surface escapes the influence of the wind, which generally rushes down from the North and up from the South poles as cold dense air. At the equator winds rise and eventually return to the poles. In their travels they touch upon the most remote portions of the globe; even the mud of the ocean floors contains windborne particles, and bits of snow are widely distributed by wind force across the vast expanses of the Antarctic Ice Sheets, an area where no liquid water exists. Contrary to the mistaken assertion that the wind lacks sufficient quantities of energy to ameliorate our energy crisis, it has a huge energy potential, a fact being verified by many organizations and men of science. Guyford Stever, the director of the National Science Foundation, has testified before a House subcommittee that "while solar and geothermal energy seem to be the best outlook, wind power may become a significant factor . . . on a regional basis."

The practicality of large-scale generation of electricity through wind power is seen as a definite possibility by the National Aeronautics and Space Agency. NASA predicts that large windmills could supply as much as 10 percent of the nation's electricity. One study covering the southwest portion of the United States concluded that if the strong wind racing across Texas could be harnessed, it could supply 8 percent of the entire nation's electricity. Even more enthusiastic is Dr. William E. Heronemous of the University of Massachusetts. In a system he has recommended, a series of huge windmills could generate, by the year 2000, about 1.5 trillion kilowatt-hours of electricity each year, virtually 20 percent of U.S. needs. This is nearly as much as the total electricity presently being generated yearly in this country.

The federal government, finally beginning to suspect that it is worthwhile to harness the energy in the wind, is spending at least $30 million on a five-year study program. Meteorologists have calculated that the power in the wind amounts to a fantastic 3×10^{17} kilowatts (i.e., 3 followed by seventeen zeros). Much of this wind is located in the upper atmosphere and is not readily available, but a lot is within reach. The amount we can harness comes out to 2×10^{10} KW or 20 billion kilowatts—or forty times the United States' present electrical generating capacity of .5 billion kilowatts.

BEFORE THE PYRAMIDS

Even before men learned the basic rudiments of writing, the sails of merchant ships were silhouetted against the setting sun on the Euphrates River in Sumeria. Men had observed the power of an invisible force, had felt it against their backs and had seen it bend the tall palm trees. They had been terrorized by its fury during destructive storms. If this force could bend a steadfast tree and even uproot it, could it not move a man-made object—a raft, for example? And so, the first application of wind power was a crude raft with a square sail. Once men became aware of the working power of the wind, they quickly took advantage of this free energy and built ships of every type, from Phoenician trading vessels, Roman war galleons and Greek cargo ships to Scandinavian longboats, Yankee clipper ships and modern-day pleasure sailboats. For thousands of years men traveled the seas without polluting the air or the waters.

Around the seventh century the idea of using the wind to perform other types of tasks took shape. The first known windmills were in Seistan, on the border between Persia and Afghanistan. But these original wind machines looked nothing like the famous Dutch windmills. They consisted of a wooden structure in which was mounted a vertical shaft running from the ground to the roof (see Figure 44). Mounted on the lower portion of the shaft were a number of sails or wooden paddles which caught the wind and rotated the vertical shaft. Openings on the sides of the building allowed wind to enter and leave the structure. On the vertical shaft, above the windmill sails, was mounted a pair of millstones for grinding wheat. This was a brute-force, direct-drive method with no gears for gaining extra power or changing direction of motion or any other controls. But it was a big improvement over manual labor, and the wind was free. Sometime later the builders of windmills, realizing that the wind was stronger the further up one went, moved the windmill sails from below the millstones to the top of the structure. These types of devices, which are called horizontal windmills, eventually found their way to China, Western Europe and even the United States.

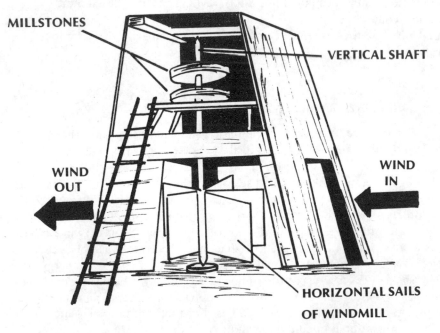

FIGURE 44. Ancient version of a windmill (644 A.D.) used to grind wheat.

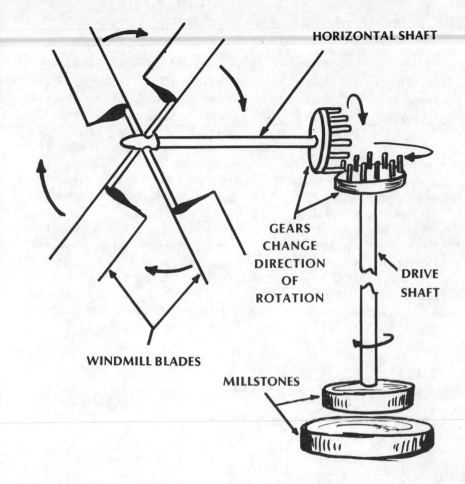

HORIZONTAL SHAFT

GEARS
CHANGE
DIRECTION
OF
ROTATION

DRIVE
SHAFT

WINDMILL BLADES

MILLSTONES

FIGURE 45. Later windmills had vertical blades with gears to change direction of shaft to drive millstones.

Over the centuries the concept of the windmill changed as it moved to the Arab countries and to Europe. Around 1180 a new version of the windmill called a post mill appeared in France. Instead of having the windmill sails rotate horizontally on the vertical shaft, the new windmill turned the sails on end (i.e., in a vertical position) and mounted them on a horizontal shaft (see Figure 45). The main advantage of this system is that vertical blades extract much more energy from the wind than horizontal blades because the blade area facing the wind is much larger. But this method has a disadvantage. In the old horizontal system the mill blades are effectively always facing the wind. The vertical-blade windmill effectively faces the wind in only one direction. If the wind changes its direction of flow, the blades will stop rotating. Therefore this type of windmill had to have blades that could be rotated to move into the wind as the wind shifted. The post mill was built to do just this. The entire mill—blades, horizontal shaft and gears, even the millstones—is included in a structure that is supported on a center post. Thus the entire upper portion of the windmill can be rotated to keep the blades facing into the wind (see Figure 46). A mill hand entered the upper portion of the mill by a ladder. This type of windmill, although an improvement on the ancient horizontal type, suffered one main disadvantage: in order to face the blades into the wind, the entire upper structure, which was quite heavy, had to be rotated. But this problem was soon solved.

At the beginning of the 14th century a new type of mill appeared on the scene; it was called the tower mill (see Figure 47). It got its name from the fact that the main or lower portion was built like a tall tower, either of stone or heavy timber. Someone had thought of the idea of simply moving the heavy millstones and gearing from the top to the bottom of the mill. A small structure called a cap, which was movable, held the windmill blades;

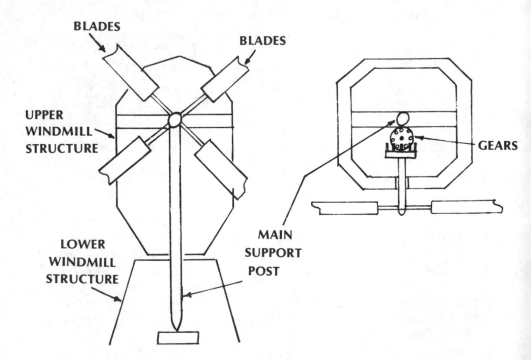

FIGURE 46. Vertical-blade windmills are built so that the upper structure can be rotated to keep the blades facing into the wind.

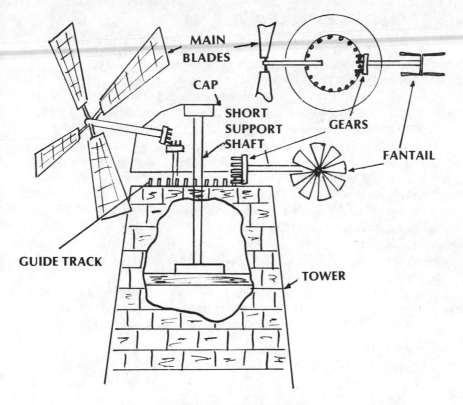

FIGURE 47. The tower mill and fantail were big improvements over older wind-mill designs.

only the small cap section at the top of the tower, and not the entire mill, had to be rotated to face the blades into the wind.

By the 1700s the windmill had been adopted in all parts of the world. Because of its extensive use, it was constantly developed and improved. An important improvement was the fantail, invented by Edmund Lee in England around 1745. Made up of five to eight blades, the fantail was mounted at right angles to the main-blade assembly, as seen in Figure 47. Basically, this device was used to automatically position the main windmill blades directly into the wind. The idea behind it is quite simple. When the wind changes direction, it begins to strike the side of the large blades; as a result, they come to a stop. However, since the fantail is facing the side perpendicular to the main blades, it is faced directly into the wind and begins to spin. The fantail is connected, through gears, to a large guide rail around the top of the tower. The force generated by the fantail rotates the top or cap of the windmill and repositions the main blades so they are facing into the wind. This was one of the earliest forms of automatic controls.

In the late 17th and early 18th centuries the Dutch built a large number of mills on the shores of Manhattan Island and along the cliff-lined coast of New Jersey. The English later built mills along the Atlantic. The last windmill in New York City burned down in the 1940s, but two of the most powerful Dutch windmills ever built are still standing on a cliff in San Francisco's Golden Gate Park.

Windmills were large, expensive structures which could not be mass produced, so not many people could afford them. But in 1854 a young mechanic from Connecticut, Daniel Holladay, designed and built a small, rugged, inexpensive windmill that would become a landmark in rural America (see Figure 48). It consisted of a series of as many as 20 steel blades positioned in a circle, like a pinwheel. Protruding from the back, perpendicular to the blade assembly, was a shaft behind which was mounted a tail vane, which kept the main blades facing directly into the wind.

This practical windmill caught on, and it was not long before factories throughout the country began to produce the Holladay windmill. At first they sold them by the dozens and then by the hundreds. Farms, planta-

FIGURE 48. In 1854, Daniel Holladay designed an inexpensive windmill that was to become a landmark of rural America.

tions, ranches and other rural communities put the mills to good use. Some were used to supply running water in homes. Cattlemen and farmers installed them for irrigation. The Union Pacific even used the Holladay windmill to pump water into trackside tanks for passing locomotives. Production of these wind-driven beasts of labor eventually was in the millions. But with the advent of rural electricity, the windmill industry came to an abrupt end. However, some windmills continued to pump: the New Mexico State University of Agriculture estimates that there are about 175,000 water-pumping windmills left in the United States, half of which may still be in good working condition.

WINDMILLS GO MODERN

As it turned out, windmills were able to keep up with modern developments. Man wanted to use electricity to light his home and operate handy appliances. Fine! Windmills could still provide an important service by generating electricity. In 1890 an inventor named P. La Cour, who lived in Denmark, built a windmill that was connected to an electrical generator, and it worked. The concept spread quickly to the United States; by the turn of the century thousands of small wind-powered electrical generators were whirling on small homesteads and farms throughout the Midwest, the South and the West. Eventually these power units came up against competition in the form of gasoline and diesel-engine generators and rural transmission lines erected by the Tennessee Valley Association. But the windmill generators gave their competitors a run for their money; as late as 1950 there were about 50,000 small windmills converting the flow of air into electricity.

Still, it seemed wind power would never get the opportunity to prove itself as a large-scale source of electrical energy until a man named Palmer Cosslett Putnam came on the scene. He had great insight into the power of natural forces and believed that the wind could supply man with much of the energy he needed. According to Putnam in his book *Power from the Wind*, "In 1934 I had built a house on Cape Cod and had found that both the winds and the electric rates were surprisingly high. It occurred to me that a windmill to generate alternating current might reduce the power bill. . . ." Five years later things began to happen. The S. Morgan Smith Company, which built hydraulic turbines, decided to look into the manufacturing of large wind turbines as one way to diversify their product line. Very few people knew much about the wind in those days, to say nothing of how to build huge wind-driven generators. This is where Putnam, a well-known consultant engineer with an adventurous spirit, came in. His proposed design for a wind-powered generator was accepted, and research and development work started shortly afterward. Actual construction began in 1940, and erection was completed in August 1941. Then followed some months of system testing. Putnam describes the moment of truth when the giant windmill went into commercial operation.

"Finally, on the night of Sunday, October 19, 1941, in the presence of the top management of the Central Vermont Public Service Corporation and many of the Staff of the S. Morgan Smith Company, with the Smiths listening in by long distance telephone, Bagley, having completed his adjustments and made his final inspection aloft, phased-in the unit to the lines of the utility company, in a gusty 25-mile wind from the northeast.

"There was no difficulty. Operation was smooth. Regulation was good. After 20 minutes at 'speed no load,' the blade pitch was adjusted until output reached 700 kilowatts. For the first time anywhere, power from the wind was being fed synchronously to the high-line of a utility system."

In the following months of operation the giant electrical power windmill on Grandpa's Knob withstood winds of 115 miles per hour (not while in operation) and generated up to 1500 kilowatts of electrical power in winds of 70 miles per hour. And, except for the failure of a bearing in February 1943, all went well with the colossus of windmills until March 26, 1945. At about 4 A.M. in the morning the tower suddenly began to shake violently. Harold Perry, who was in the control room, managed to get to the controls and shut down the windmill. Outside, in the darkness, one of the 8-ton, 87-foot blades had broken off and was hurled 750 feet through the air. It landed on its tip. After extensive inspection and subsequent studies, it was decided to abandon the project. Up until the time of the accident the windmill had generated an impressive 360,000 kilowatt-hours of electricity. Now it stood there like the corpse of a one-armed giant. However, a great deal of knowledge had been gained. According to Beauchamp E. Smith of the S. Morgan Smith Company, "In six years of design and testing of the 175-foot, 1250-kilowatt experimental unit on Grandpa's Knob near Rutland, Vermont, in winds up to 115 miles per hour, we have satisfied ourselves that Putnam's ideas are practical and that regulation is sufficiently smooth. We think we could now design, with confidence, 2000-kilowatt wind-turbines incorporating important improvements leading to smoother operation, simpler maintenance, and lower cost." What failed was not the concept of wind power, but our knowledge of metallurgy: the stress on the blade shaft was simply too much. Today such a problem would not exist because of the advancements in new materials and in the science of stress analysis.

NATURE OF THE WIND

Before we commit large sums of money to multimegawatt generating stations, some basic research must be completed so that we can use the wind most efficiently. The atmosphere and its movements (winds) are highly complex in nature, and the characteristics of wind motion must be carefully considered in designing a wind-powered generator if it is to be competitive with other forms of energy generation in efficiency.

About 80 percent of all the air that surrounds our planet is contained within the region known as the troposphere. This region, which extends upwards to about ten miles, is where most of our weather originates. Wind is simply air in motion. Locally, winds can flow in all directions, changing from day to day or even hour to hour. But the earth does have some general wind patterns which are recognized because of their constancy. For example, there are the northeast trade winds and the southeast trade winds, the north polar easterlies and the south polar easterlies. There are jet streams, rivers of air that flow at speeds of 150 to 400 miles per hour at altitudes of 10,000 to 40,000 feet.

Wind is caused by the temperature differences around the earth (see Figure 49). Because the poles of the earth do not receive as much sunlight as the equator, the temperature at the poles is much lower. When air gets colder, its molecules are packed more closely together. Cold dense air becomes heavier; therefore air at the poles has a tendency to fall. Meanwhile, at the equator the air is heated by the greater intensity of sunlight. As air gets warmer, its molecules begin to spread out and a type of light air bubble forms. Like an air bubble in water, it begins to rise, leaving a partial vacuum. When this happens, the cooler air adjacent to it (higher-pressure area) moves in to fill the void. It is the movement of this air that comprises the wind. The difference in temperatures between the two areas, plus some other factors, determines the force and velocity of the wind.

There is a general rising of air at the equator, and this air moves toward the poles of the earth. At the poles there is a general falling of air, and this air moves toward the equator. According to these effects, there should be an even, steady flow of air from the poles directly to the equator. But that is not the case. A number of other actions affect the flow of the wind. For one thing, the earth is rotating constantly, and since the atmosphere does not keep up with this rotation, it has a tendency to slip or move along the direction of the equator. Actually, the wind will move from the northeast to the southwest in the northern hemisphere (the Coriolis force). The speed, direction and flow characteristics of the wind are also affected by surface friction, local thermal convection, frontal movements (a front

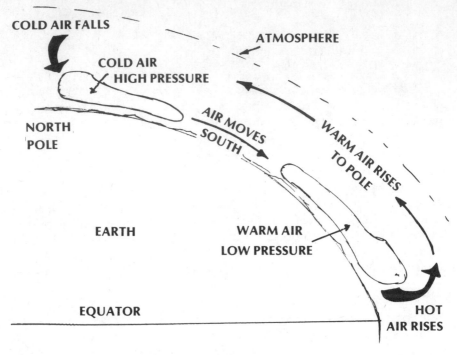

FIGURE 49. Air is sent into motion by the differences in temperature around the earth.

is a boundary between two air masses) and the earth's topography. And besides the movements of large air masses, there is activity within the masses themselves.

The structure of an air mass is highly complex. Air has a very low viscosity (ability to flow), about six times lower than water. Therefore any movement of air at speeds greater than two or three miles per hour is basically turbulent, though this turbulence is barely perceptible to someone standing in the flow. Within the mass of turbulent air, individual molecules of air move randomly in all angles as the total air mass proceeds forward. To add to the complex nature of the air mass, there is a second mode of motion called laminar flow. In the laminar flow mode air moves in a series of layers parallel to each other at different speeds. The effect of this action is that the air layers slip past each other, somewhat like new playing cards, without any mixing action taking place (see Figure 50).

All the various motions of the wind—its long distance travel and its

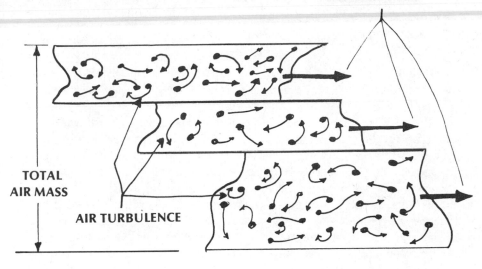

LAMINAR FLOW

TOTAL AIR MASS

AIR TURBULENCE

FIGURE 50. The structure of a moving air mass is highly complex.

local turbulence—represent a fantastic amount of energy. Energy is the ability to do work, and wind certainly does its share of work. Aside from the destruction caused by storms, the result of wind action in other areas is quite surprising.

First of all, the distance that wind will carry dust and small particles of sand is amazing. Dust picked up in North Africa will frequently be deposited 2000 miles away in England or northern Germany. At times, dust will fall on the decks of ships at sea 1000 miles from the nearest land. Australian dust will travel 2500 miles to New Zealand. But the longest trips are taken by dust that is spewed into the atmosphere by volcanoes. This dust can travel around the entire planet several times before settling back on the ground.

The second indication of the wind's work capability is the tremendous amount of material that can be carried along with an air mass. It has been estimated that some storms have transported as much as 100 million tons of dust. Consider how much fuel would be consumed by man's machines to move such a huge amount of material. An excellent example of the wind's ability to accomplish a great deal of work can be seen in a troughlike excavation west of Laramie, Wyoming, called Big Hollow. It measures 9 miles long and 150 feet deep. In order to form this closed depression, the wind had to lift and transport about 10 billion tons of dust.

MODERN WIND SYSTEMS

What constitutes a good wind-powered generating system? The answer to this question can be divided into three areas: the site location, the generator, and the blade and control assembly.

First, the site location. Obviously, one would look for a site with a lot of wind. But the presence of wind is not the only requirement; its flow is also important. An area may have very strong prevailing winds but its topography may be wrong. Hills or excessively tall trees or man-made structures may cause the wind to veer or tumble, resulting in turbulence that reduces its energy potential. So a windmill should be located in an area with a relatively smooth, strong air flow that is fairly consistent in average magnitude. Where are the areas of high wind energy? Generally, in the New England states (especially Mount Washington, which has recorded winds up to 200 miles per hour), the Great Lakes area, the Great Plains, the Pacific Coast, Alaska, Hawaii and Texas.

The speed or velocity of the wind is also critical. The amount of energy a windmill can extract is not *directly* proportional to the speed of the wind: if the velocity of the wind doubles, the power output of a windmill does not double, it increases a lot more. The output power of a windmill increases with the cube of the wind velocity. Let's say that at 2 mph of wind velocity, a windmill generator produces 10 watts of electrical power; if wind velocity doubles to 4 mph, the windmill will generate 80 watts of power ($2 \times 2 \times 2 \times 10$). If wind speed triples to 6 mph, the output of the windpower generator would be 27 times greater—$3 \times 3 \times 3 \times 10$, or 270 watts. This is why it is so important to locate a wind-powered generator on a site with high-velocity winds.

Another aspect to consider is the height of the tower upon which the blades are mounted. Generally speaking, the higher the better, because wind velocity is increased with altitude and height minimizes the effects of ground turbulence. However, the height of the tower is limited by economics. Above a certain height, the cost of the tower can become excessive and render the entire power-generating system uneconomical. The National Science Foundation is presently studying a 1000-kilowatt generator mounted on a 17-story (170-foot) tower. Of course, if the generating system is very large, the cost of a very high tower can be more readily absorbed into the overall system cost. The crucial point is that the blades of the windmill must be clear of any obstruction that could spoil the wind flow directed at them. Even in a small system there should be no line-of-sight wind interference for at least 400 yards. This is especially important because even under ideal wind conditions, a perfectly efficient windmill can extract a maximum of 59 percent of the energy from the wind passing through its blades. Since man's best designs are less than ideal, the amount of wind energy available to a power plant will be much less than 59 percent. It will, in fact, be closer to 33 percent.

Once the windmill has been located in the ideal position, the next step is to ensure that the mechanism itself is of the best design possible. One important item to be considered is the part that actually produces the electricity—the generator. As explained previously, an electrical generator is a device that produces electricity by moving a conductor through a magnetic field. In conventional generators a center portion called a rotor or armature rotates inside another part called the field-winding assembly. The key point is that the generator must be rotated. Historically, on an electric windmill the generator has been located at the top of the tower, although in some instances it has been mounted at the bottom of the tower. There are pros and cons to both of these systems. If the generator is placed at the top, the added weight (of the generator plus its gearing mechanism) requires a stronger tower. On the other hand, if the generator is placed at the bottom, special bevel gears would be required to change the turning motion from the horizontal direction of the windshaft to a vertical direction in order to drive the generator at the bottom. The use of the bevel gears plus the long vertical drive shaft would diminish the power output of the system.

Another important aspect of the generator's operation is its speed of rotation. Although automotive-type generators have been used in windmills, they are not very suitable because they are extremely inefficient at speeds under 1800 revolutions per minute (RPM) and a windmill rarely spins faster than 300 RPM. Furthermore, the generator must be constructed to produce 60 cycles per second if it is to provide power for direct use in homes or factories. One solution to the speed problem is to use gears. Just as gears are used in a car to allow it to go faster or slower, a set of gears can be placed between the windmill blades and the generator. These gears will increase the speed from the blades so that they will rotate the generator at a much higher speed.

Another way of getting around the generator speed problem is to redesign the generator, usually by making it a lot larger in diameter so that it will produce electrical power at much lower speeds. The National Aeronautics and Space Administration is presently experimenting with large-diameter generators in hopes of increasing the efficiency of wind-power systems. One reason NASA is trying this approach is that gears create friction and friction robs the system of power. Gears also require lubrication and are subject to failure. The fewer the moving parts in the system, the better and more reliable it will be.

A unique approach to the generator speed problem was taken by

Professor Macel Jufer (a Swiss generator designer), Walter Schoenball (team leader), Jacques Dufournaud (a French electronics designer) and Hans Goslick (a German blade expert). They attacked the problem in two ways. First, they designed a large-diameter generator to provide maximum efficiency. Now, the size of the generator, mounted at the top of the tower, can affect the wind flow—that is, a very large diameter unit can act as a wind spoiler and cause air turbulence in the area of the blades (incidentally, the tower structure itself can affect the performance of the blades). Therefore there are practical limits to the size of the generator. The team partially solved the speed problem by using two sets of windmill blades that

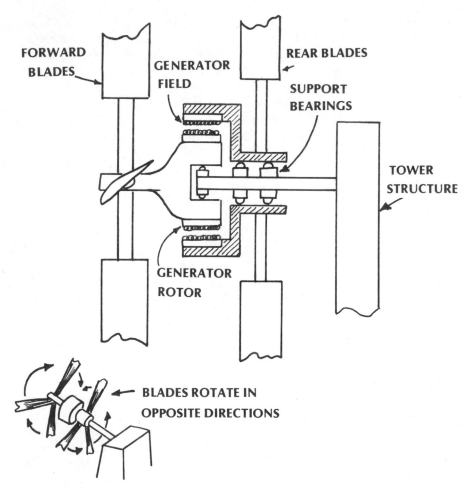

FIGURE 51. Contraroating blades and large-diameter generators eliminate the need for gears to increase the speed of the windshaft.

revolve in opposite directions. These are called contrarotating blades. As shown in Figure 51, one set of blades is attached to the forward shaft, which rotates the rotor portion of the electrical generator. The second set of blades is connected to the rear shaft, which rotates the stator or field windings of the generator. Since the curve or airfoil of the rear blades is at an opposite angle to that of the forward blades, the two sets of blades rotate in opposite directions. Since the blades move in opposite directions, the stator and rotor of the generator will also rotate in opposite directions, thereby doubling their relative speed (the speed at which they rotate past each other). In effect, the generator rotates at twice the speed of the windmill blades without using gears.

Another consideration is generator efficiency. Since the object is to extract as much energy as possible from the wind, generator design is critical. The finest conducting, magnetic and insulation materials must be used. The technology that can be extremely valuable in wind power (as well as in other methods of electrical power generation) is cryogenics.

As described previously, a superconducting generator (one that operates under extremely low temperatures) can produce two or three times as much electricity as a conventional generator. If the generator in a regular wind-power system designed to produce 100 KW were replaced with a superconducting generator, that same system, with the same amount of wind velocity, could generate 200 or even 300 KW. But the generator can do its work only if it is driven with sufficient force, and that is the job of the blades.

The blades are the most critical part of a windmill system: their task is to directly extract energy from the wind. Since they are subject to numerous forces, including wind pressure, centrifugal force and vibration, both their shape (configuration) and the material of which they are constructed must be carefully considered. In the pioneering wind project on Grandpa's Knob the *system* worked well, it was the blades that failed. Modern designers of windmills are carefully reviewing the information available from the aircraft and helicopter propeller industry for clues to high-efficiency windmill blades. Their goal is to design blades that are better than the fat, slow-moving, clumsy, inefficient arms used on old windmills. The new emerging blades are long, slender and engineered to spin as fast as possible with the smallest amount of wind.

Scientists are also trying to learn a lot more about the forces acting on the blades, including the interaction between the tower and the blades. For example, the blocking of the wind by the tower may cause the blades to flap. This is explained by the fact that each time the blade passes the tower, the effective air pressure on it may lessen, causing the blade to assume a slightly different position (in its horizontal axis). As it rotates free of the tower, the regular wind pressure may resume, pushing the blade back to its former position. This would occur each time the blade passes the tower, resulting in a flapping action. Such an action, if it occurred over a long period of time, could put additional stress on the blade and lead to metal fatigue followed by blade failure.

There are three basic requirements for blade materials: strength, ease of fabrication and potential low cost. Because generators are more efficient at high speeds, it is best to have the windmill blades rotate as fast as possible. But high speed exerts a great amount of centrifugal force on the blades (tending to tear the blades off the shaft), especially at their tips, which travel at the greatest speed. Consequently, there must be a compromise between ideal blade speed and ideal blade materials. The strength of the materials obviously determines maximum blade speed. Presently a host of materials are being tried on a number of types of blades. Some use steel alloys or aluminum, while others use fiberglass and even cloth.

In order to get around the problem of extreme centrifugal force, Wayne Wiesner, manager of the wind energy program at Boeing Vertol, is considering going to a totally different material—fiberglass. He would like to build blades up to 150 feet long. Fiberglass, explains Wiesner, is stronger than metal and can better withstand blade-tip forces. With the longer blades, the tips can reach a speed of 200 mph (the giant arms of the old Dutch mills rotated a maximum of 80 mph). At this speed, the blades could run a 1000-KW generator in winds averaging 20 mph (which are not unusual in many areas of the United States).

Although not basically new, the technical refinements of the new blade designs are fascinating. The original windmill blades were of two basic types: the horizontal paddlewheel blade that rotated on a vertical shaft (Figure 51) and the later vertically oriented set of arms that turned around a horizontal windshaft (Figure 45). In the United States the multiblade vertical windmill was developed (Figure 48), and later an aircraft-type propeller was used on the modern wind generating system on Grandpa's Knob. The latter type of blade comes in a number of configurations —two blades, three blades and more. In fact, in a prototype system on the German North Sea island of Sylt, a 55-KW wind generator uses two sets of five-blade propellers (contrarotating blades). There are advantages and disadvantages to all these approaches. A great deal of research is presently underway to determine the best combination of blades.

A number of different approaches to blade design appear quite promising. One is called the bicycle-wheel wind turbine. As its name implies, its structure is similar to that of a bicycle wheel. This modern approach to windmill blade design is the work of Tom Chalk, who runs a small factory near St. Cloud, Florida. His blade assembly consists of a bicyclelike outer rim and a smaller inner rim with 48 blades mounted between the rims (see Figure 52). Steel wires, like bike wheel spokes, extend from the center shaft to the rims and provide rigidity to the entire blade assembly. The outer rim, which supports the 48 blade tips, prevents them from vibrating like unsupported blades. Dr. William Hughes of Oklahoma State University (which has been conducting wind power research for many years) is very enthusiastic about Chalk's design, referring to it as a work of "near genius." Aside from being very sturdy, the bicycle-wheel wind turbine is highly efficient. Its efficiency is the result of a number of factors. First, reduced blade vibration, due to the outer support, precludes a loss of efficiency from vibration. Second, the entire assembly is very light. And finally, the assembly has a large blade surface area (48 blades). Besides being proportional to the cube of the wind speed, the efficiency is also proportional to the square of the blade surface area. Dr. Hughes plans to

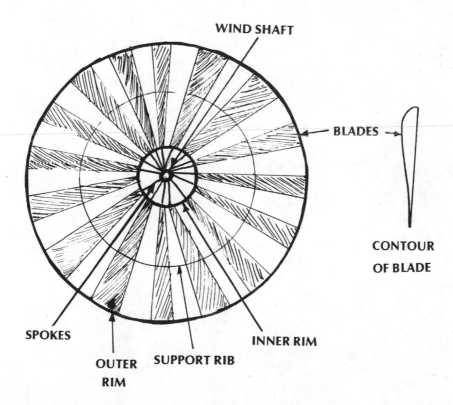

WIND SHAFT

BLADES

CONTOUR
OF BLADE

SPOKES

INNER RIM

OUTER
RIM

SUPPORT RIB

FIGURE 52. The bicycle-wheel wind turbine is highly efficient and minimizes blade vibration.

combine the advantages of Chalk's turbine with a special constant-frequency generator he has developed and eventually design a 150-foot bicycle-wheel wind turbine system that will produce electrical power in the megawatt class.

At Princeton University engineers are taking a different approach to wind power. Throughout the blustery winter of 1973 and 1974 a candy-striped pole with an odd-looking propeller spinning at its top stood on Princeton's campus. It endured harsh winds and proved that the Sweeney Sailwing generator can meet the challenge of wind power. This new type of windmill blade, designed by Thomas E. Sweeney, director of the Advanced Flight Concepts Laboratory at Princeton, is an offshoot of a high-performance aircraft-wing design. Sweeney, whose specialty is low-speed aerodynamics, originally designed the sailwing to get maximum lift with minimum

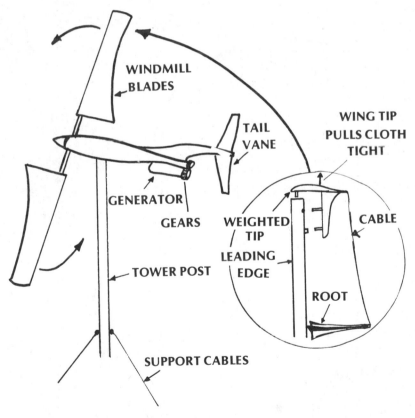

FIGURE 53. The sailwing offers the advantages of light weight and durability.

weight. NASA tested his sailwing (which is half the weight of a conventional wing) in its Langley Research Center wind tunnel. It worked well. If it worked as an airplane wing, why not as a blade, which is also an airfoil?

Sweeney explained, "While working on the sailwing aircraft, I became intrigued with the idea that the sailwing might make an efficient windmill blade." And so he set to work to prove his theory. In 1968, long before the energy crisis, Sweeney's group built a 10-foot-diameter sailwing windmill. This was followed by the construction and testing of a 25-foot-diameter structure, which is the best size for research purposes. It is convenient to work with, economical to build and large enough to provide accurate, realistic data. The sailwing's light weight and sturdiness are due to its unique construction (see Figure 53).

The blade is actually hollow. Its leading edge is a separate tubelike structure. At the outer edge is located a weighted tip, which fits into the leading edge but is not anchored securely, that is, it can move slightly. On the other end of the leading edge structure is a root fitting that is rigidly connected. A cable is connected between the tip fitting and the root fitting. The entire blade structure is covered with Dacron (the material for boat sails). As the blades begin to spin in the wind, they assume an ideal airfoil shape. Right now the Sweeney Sailwing is as efficient as the best wind turbine ever built. It has a lifting or air drawing power (lift-to-drag ratio) 20 times greater than the drag it creates. So it develops a great amount of power for its weight, which is only 44 pounds for the 25-foot-diameter sailwing. That is only a half-pound per square foot (lightweight aircraft wings are two and a half times heavier). Because of the problem of centrifugal force, as previously explained, weight is a critical factor in windmill blades. Yet the sailwing's light weight does not diminish its strength; it has survived winds of 160 miles per hour in wind-tunnel tests. The Sweeney Sailwing design is so promising that the Grumman Aircraft Corporation has acquired a license to develop and market small wind generating systems using it.

Another wind-powered generator design, the catenary, bears practically no resemblance to a windmill. A catenary is actually a type of curve that results when a free-hanging flexible cable supported at its ends is rapidly rotated. The catenary looks as strange as its name. It consists of two or three ribbonlike curved strips mounted on a vertical shaft, (see Figure 54). Also known as the Darrieus wind turbine, this odd-looking machine was developed by the French inventor G. J. M. Darrieus in 1925. Unfortunately, the Darrieus or catenary turbine suffers one major disadvantage: at very high wind speeds it loses almost all its power-generating ability (its angle of attack and lift-drag ratio fall to zero); and at very low

This 100 kilowatt windmill being built by NASA's Lewis Research Center, Cleveland, is a forerunner of much larger systems which may supply up to 10 percent of the nation's electricity by the year 2000. (Courtesy NASA)

speeds (below 10 mph) the blades stall and, again, it loses its power-generating ability. Therefore additional devices must be added to the vertical shaft to aid the turbine to start up in wind from any direction. But the catenary windmill enjoys a number of important advantages. One, it is simple in design and can be built cheaply using readily available materials. Two, because of its vertical orientation, it can accept wind from any direction and therefore does not need expensive mechanisms to keep it pointed at the wind. Nor does it lose power momentarily as the wind changes direction. Three, the catenary wind turbine does not require a large tower to keep its long blades clear of the ground. Four, the generator can be located at ground level, which precludes the need for a heavy support structure. In a catenary wind turbine the lower ends of the blades are connected to a short vertical shaft that drives the electrical generator.

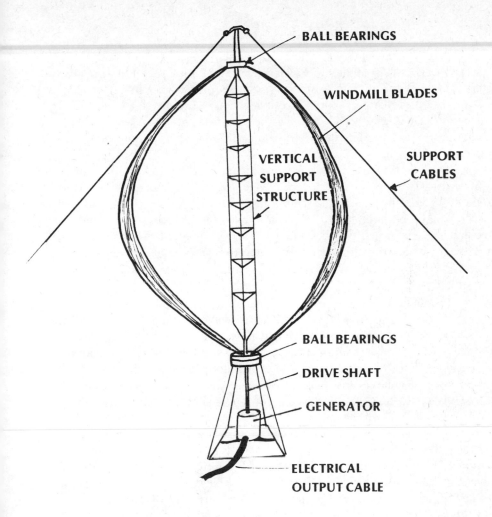

BALL BEARINGS

WINDMILL BLADES

SUPPORT
CABLES

VERTICAL
SUPPORT
STRUCTURE

BALL BEARINGS

DRIVE SHAFT

GENERATOR

ELECTRICAL
OUTPUT CABLE

FIGURE 54. The vertical-axis catenary wind turbine can operate efficiently no matter which way the wind is blowing.

A catenary wind turbine system has been designed by two Canadian inventors, Raj Ranji and Peter South of the National Research Council in Ottawa. Their test model, which stands about 20 feet high, can generate about 1000 watts of power in a 15-mph wind. This type of windmill can operate in winds as low as 10 mph.

ERDA has taken a deep interest in the catenary wind turbine. A 15-foot vertical-axis three-bladed wind turbine is presently undergoing tests at its Sandia Laboratories near Albuquerque, New Mexico. ERDA feels that this type of device is significantly cheaper than conventional windmills—as much as seven times less. The National Science Foundation (NSF) awarded about $3 million for wind energy studies in 1975, mainly to aerospace companies which have a great deal of expertise in aerodynamics and generator design. General confidence is building. As explained by Louis V. Divone of NSF, "There is no question that they [windmills] are technically feasible. The question is how to design them according to modern technology."

The country hasn't done any significant research in wind systems for 30 years. There are three real challenges when it comes to wind power. First, there is the windmill assembly itself with all its associated problems, which have been discussed. Then there is system cost, which must be competitive with that of other energy systems. Finally, there is the problem of what to do with the electricity generated—can it be fed directly into power grids, can it be stored, etc.? Let us now take a look at the cost problem.

Production costs are presently high for wind systems and will have to be brought down. A wind-powered electrical generating system that can compete in cost with coal and nuclear systems must be developed. But once wind systems are mass-produced, costs will fall. In fact, we may have already begun to reach that point. According to Robert K. Swanson of the Southwest Research Institute, large wind-powered electrical generating systems could now be constructed for approximately $350 to $400 per kilowatt-hour (kwh) capacity. Gas-fired steam turbines currently cost $250 (and costs will rise sharply), and conventional atomic power plants cost $400. Coal-fired plants with air pollution equipment also run $400. Louis Divone feels that competitive wind-powered systems are well within reach. Many other prominent experts in the field agree with him—professor William Heronemous, for one. As he explains, "By 1979, wind-powered systems will cost less than nuclear plants—and be completely safe." In fact, Professor Karl Bergey of the University of Oklahoma has designed a wind-powered generation system that can produce 150,000 kwh annually, with the electricity costing only 2 cents per kwh. The contrarotating wind turbine system on the island of Sylt is already producing electricity at a cost that is competitive with conventional systems.

The cost challenge can definitely be met, with a few extra bonuses thrown in for good measure, such as an absence of pollution, freedom from hazards and an inexhaustible supply of free fuel. So the next thing to consider is electricity use. Where will we locate wind systems? How many will there be? How big will they be? How will they fit in with our existing electrical grid systems?

Wind-powered generating systems will fall into three basic categories. The first is local use of a small windmill. Thousands of such small systems were in operation in remote areas of the country during the early part of this century, and surprisingly, windmill power is becoming popular once again in rural areas. Many a modern pioneer is opting for his own independent source of electricity.

The average household uses between 400 and 500 kwh of electricity each month, not including the power needed for water heaters, clothes dryers, toasters and other high-demand items. To provide electricity for a refrigerator, lights, stove and small appliances, a wind turbine with a 20-foot propeller, capable of generating 8 kw (assuming a 10-mph average wind), would be required. Figure 55 shows what a typical small wind-powered system would include.

The windmill itself should be placed on a tower approximately 40 feet high (the height will vary with terrain, natural and man-made obstructions,

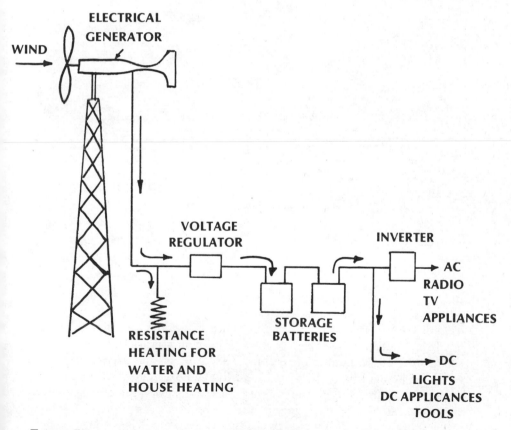

FIGURE 55. A typical system for household electrical needs in rural areas.

Vertical-axis windmills are being studied by NASA's Langley Research Center, Hampton, Virginia, for future electric power applications. (Courtesy NASA)

etc.). The electrical output can be used directly to energize resistance-type units such as hot-water and baseboard heaters. Electrical power from the wind-driven generator would flow through a control device called a voltage regulator (similar to the ones on cars), and into storage batteries, which are required for a number of reasons. They store all the excess electricity generated by the windmill and not used by the household appliances. And since the wind frequently changes in speed and intensity, the electrical level also changes; as a result, lights in the house using this electrical source would dim and glow brighter with changes in the wind. The voltage regulator prevents this from happening by passing only electricity above a certain voltage to the batteries. Once the electrical voltage falls below a certain level, the voltage regulator isolates the wind generator from the storage batteries.

The number of batteries required varies according to how much electrical power is to be stored. A household can require up to 20 heavy-duty batteries. Storage batteries for this application are built with heavier plates and are capable of withstanding the stresses of as many as 2000 recharges. Battery storage capacity should be sufficient to supply electricity for a household for four to seven days.

It is also possible to take advantage of a very interesting storage facility—the local electrical power utility's system. A household wind generator can be hooked into the output portion of the electrical meter. With such a hookup, extra electrical power from the windmill can be sent back through the meter, thus running the meter in reverse and cutting down on electricity bills. In this way, one would be using the utility's system as a giant storage battery. But what about the television, hi-fi and other similar appliances that cannot operate on DC power but require 60-cycle AC electricity? Here is where a device called an inverter comes in. An inverter changes DC power into AC power. And that is about the extent of a small wind-driven electrical generating system.

Private wind systems may be fine for those who can afford them and do not mind the work of maintaining them. But these people are only a small percentage of the population, so such systems will not affect the overall energy crisis to any meaningful extent. What is needed are large wind-power systems that can generate megawatts of power. If by 1980 wind power were to be called on to generate just 1 percent of the United States' electrical requirements, it would have to produce more than 650 MW of power. This would require the construction of 325 200-foot propeller windmills. Windmills of this size are not unrealistic—there are no technological problems—but they are complex and require advanced electronic controls for a number of functions. The propellers, for example, must be carefully monitored to ensure optimum pitch. As the strength of the wind changes, the pitch or angle of attack of the blades must be changed for maximum efficiency. In the event of an extremely strong wind (during a hurricane) the propeller blades may have to be feathered (their leading edge pointed directly at the wind) to prevent overspeeding and damage to the system.

Electronics must also be used to keep the windmill pointed in the direction of the wind. It takes both tremendous power and precision to quickly rotate a massive structure with a spinning 200-foot propeller. Finally, the problem of storing large amounts of electricity generated by windmills during off-peak hours must be resolved. But a number of methods (previously discussed) have been proposed for this purpose. As shown in Figure 56, electricity from wind-driven generators need not be used

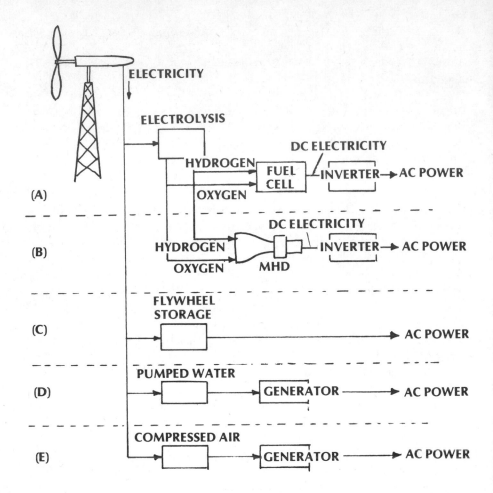

FIGURE 56. Electricity from wind-driven generators can be stored in a number of ways.

directly. After all, the output of the generator will vary with wind speed. The generated electricity can either be converted into another form and be used to run another type of constant power generator, or it can be stored for later use during a peak demand period. For example, the electrical power can be used to operate an electrolysis system to produce hydrogen and oxygen; these two chemicals can then be used to operate a fuel cell (Figure 56A), which produces a steady level of power. Or the hydrogen and oxygen could be used to run an MHD (magnetohydrodynamics) generator (Figure 56B). In both cases an inverter would be needed to change the electricity from DC to AC. The windmill power could also be used to spin a huge flywheel, pump water up to a storage reservoir or pump compressed air into a large chamber (Figure 56C, D and E). In each of these cases the energy would be available to operate a conventional electrical generator when the electricity was needed. So the criticisms raised about regularity of power and storage are not as serious as one might think. In fact, compared to the more familiar forms of power generation such as coal-fired steam generators and atomic power, wind power is relatively problem-free.

There have also been proposals for connecting a large number of giant windmills directly into a national or regional power grid. The windmills would be designed to produce constant-frequency (60 cycles per second) electricity. Since the windmills would be distributed over a large geographical area, a number of them would be feeding power into the grid at any given time. If there were an excess amount of power, it would be stored. If their output fell below a specified minimum, supplementary power from another system (solar, geothermal, tidal, solid waste or conventional fossil fuel) would bring the power level back to normal. Professor Heronemous envisions a chain of windmills stretching south to north, from the great plains of Texas to the Canadian border. To conserve land, 800-foot towers would be built every half-mile, straddling the highways with heavy cables connecting them. The windmills would be slung on these cables.

Is this a dream? Can the wind really supply meaningful amounts of power? In a special report (NSF/RA/N–73–001) by the National Science Foundation it was estimated that the winds of the Great Plains, valleys, continental coastal shelves and the Aleutian arc have a power potential of 100 million MW. This greatly exceeds the total quantity of electricity we will need in 1980, 1990 or 2000. The accompanying table is a partial list of the distribution of wind power in the United States. Certainly such a vast source of clean energy deserves our serious consideration. The National Science Foundation thinks it does; it spent $865,000 in 1974 and 1975 on a 100-KW experimental wind-powered generator designed and built by the Lewis Research Center in Sandusky, Ohio. NASA is also building and testing two smaller systems. One will be in the range of 50 to 250 KW. The other will be capable of producing up to 3 MW of power. According to Ronald L. Thomas, project manager at Lewis, the windmills will be completed and fully operational by the fall of 1976. The overall plan is to have six wind-driven electrical generators delivering electrical power in various locations of the United States by 1979. The NSF windmall will be of a 125-foot-diameter, two-bladed, automatic-pitch-control design. It will stand atop a 125-foot tower and will be able to generate electricity in breezes as low as 7 mph. A final evaluation report on the effectiveness of the system is due in 1978. This will be followed by a 1-MW generator.

MAXIMUM ELECTRICAL ENERGY PRODUCTION FROM WIND POWER
(Source: NSF/RA/N–73–001.)

SITE	ANNUAL POWER PRODUCTION	MAXIMUM POSSIBLE BY YEAR
(1) Offshore, New England	159×10^9 kwh	1990
(2) Offshore, New England	318×10^9 kwh	2000
(3) Offshore, Eastern Seaboard, along the 100-meter contour, Ambrose shipping channel south to Charleston, S.C.	283×10^9 kwh	2000
(4) Along the E-W Axis, Lake Superior (320 m)	35×10^9 kwh	2000
(5) Along the N-S Axis, Lake Michigan (220 m)	29×10^9 kwh	2000
(6) Along the N-S Axis, Lake Huron (160 m)	23×10^9 kwh	2000
(7) Along the W-E Axis, Lake Erie (200 m)	23×10^9 kwh	2000
(8) Along the W-E Axis, Lake Ontario (160 m)	23×10^9 kwh	2000
(9) Through the Great Plains from Dallas, Texas, north in a path 300 miles wide W-E, and 1300 miles long, S to N. Wind Stations to be clustered in groups of 165, at least 60 miles between groups (sparse coverage)	210×10^9 kwh	2000
(10) Offshore the Texas Gulf Coast, along a length of 400 miles from the Mexican border, eastward, along the 100-meter contour	190×10^9 kwh	2000
(11) Along the Aleutian Chain, 1260 miles, on transects each 35 miles long, spaced at 60-mile intervals, between 100-meter contours. Hydrogen is to be liquefied and transported to California by tanker.	402×10^9 kwh	2000

Estimated Total Production Possible: 1.536×10^{12} kwh by year 2000

There is a lot of activity in the area of wind power because there is a lot of power in the wind. The need exists and the conditions are becoming right. The demand for energy in general and electricity in particular is rising. But also rising are the prices of oil, coal and gas (in some states natural gas is already over a dollar per 1000 cubic feet and going up.) Wind power's time has come. Certainly, a lot of research and development must still be done. To draw an analogy with the aerospace industry, we are still in the V-2 rocket stage. But it is just a matter of time and effort before these machines begin to contribute important amounts of energy to our country. The 200-KW units will be first, followed by the 2000-KW units, and the 1980s or 1990s may see units as large as 10 MW. Once we start connecting thousands of these units into the nation's electrical grid system, the contribution will be quite significant in terms of available energy, savings in fossil fuel costs and above all reduced pollution.

We should think of windmills as vehicles that can reach into the sky and draw from nature's inexhaustible supply of clean energy. Converted into hydrogen and oxygen, this could be an excellent form of energy for many sectors and activities of our society. The important thing to recognize is that the energy is there, just waiting to be used.

8

Tapping the Earth's Blazing Core

William Bell Elliott was an explorer and land surveyor who spent much of his time roaming through quiet, unsettled forest lands, rolling hill country and mountain regions. About two years before gold transformed San Francisco into a lawless boom town, he was making his way through the mountains north of the city, around the area of Mayacmas–Clear Lake. Suddenly, as he reached the rim of a hill, he was confronted with a scene that stopped him dead in his tracks. It was something he had never seen before and it strongly suggested that he had stumbled across the very gates of hell: huge jets of steam were hissing from the earth and forming high columns of billowing clouds that dissipated into the sky.

Little did Elliott know that he had chanced upon surface evidence of the violent forces of heat ascending from the earth's blazing core—a burning core that had begun to accumulate heat billions of years before the first complex molecule began to evolve into a primordial life form. The valley into which Elliott had wandered is today known as the Geysers, and the modern term for the source of heat that generated the hellish columns of steam is geothermal energy. Today, instead of piercing the earth's surface with shovels and picks in search for gold, modern prospectors are using complex electronic equipment to find places where vast amounts of energy, in the form of heat, are located close to the surface.

Until recently, a lot of methods for generating power were viewed as "interesting" but impractical—geothermal energy being one of them. But as in the case of wind and solar power, there is renewed interest in this potential source of energy. The concept of geothermal energy invokes the image of a gigantic furnace in the center of the earth that can supply heat energy for thousands of years into the future. One could, in fact, consider the core of the earth a huge atomic reactor (a light-water reactor) that continuously leaks heat to the surface.

Geothermal energy has been known and used for over 2000 years. The Romans had special baths used for health reasons; these were naturally heated springs whose water contained significant amounts of sulphur. These luxurious baths, frequented by the wealthiest citizens, were located all around the Mediterranean, even as far north as Bath in the British Isles. Hot springs and medical spas existed in ancient Japan and in other parts of the world. But two millennia had to expire before man began to put this natural form of energy to work. Before one can clearly understand how this is accomplished, one must know a few fundamental facts about the structure of our planet.

THE STRUCTURE OF THE EARTH

The earth is divided into three basic levels (see Figure 57). The outer level, upon which we walk, is the crust; this level reaches down 4 to 5 miles where the earth's oceans lie, and is 20 to 35 miles thick at the locations of the continents. The crust is relatively soft compared to the next level, which is called the mantle. This mantle consists of very hard rock and reaches to a depth of 1800 miles, bordering on the outer edge of the earth's burning hot center, known as the core. The core is about 4300 miles in diameter and very hot; temperatures are far in excess of 10,000° F. There are a number of theories as to why the core is so hot, but in general there are two causes. First, the tremendous pressure of the earth's gravity tries to compress the trillions of tons of matter tightly together. Second, the earth contains a large amount of radioactive material including uranium 238, uranium 235, thorium 232 and potassium 40. As in an atomic reactor core, the activity of these radioactive materials generates a tremendous amount of heat. This heat tends to flow outward, but much of it is contained in the core by the surrounding mantle. However, the amount of heat that does reach the earth's surface is significant, about 150 BTUs per square mile each second. To put it simply, the heat that reaches the earth's surface in one year is about equal to the heat value of 170 billion barrels of oil.

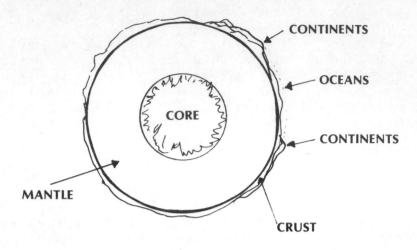

FIGURE 57. The structure of the earth comprises three different levels. Its center reaches temperatures of many thousands of degrees.

The upper layer of the earth, the crust, is actually made up of three layers (see Figure 58)—the lithosphere, asthenosphere and mesosphere—each of which has its own unique physical structure and characteristics. In general, the earth's temperature increases roughly 1° F. for every 100 feet in depth. But in some parts of the earth the local temperature of the crust (more accurately, the lithosphere and asthenosphere) is much higher; these are the hot spots of the earth and they receive their heat in a different way from the rest of the earth's surface. These hot spots range from 150° to 750° F. The cause of these hot spots will be discussed shortly. But first let us see what happens as a result of their existence.

Very high temperatures, up to 700° F., can be found almost anywhere in the United States at depths of about 20,000 feet. This heat comes from molten rock, or magma, which is heated by the earth's core. As shown in Figure 59, the magma is usually situated just below a thick layer of solid, impervious rock. While the solid rock acts as a container for the magma, it allows heat to pass through it to the next level. In areas where geothermal energy exists, this level consists of softer rock which may be porous. Just above the porous rock level lies a second level of solid (impervious) rock.

Heat from the magma passes through the lower level of solid rock and enters the porous rock, thereby raising its temperature to anywhere between 150° and 700° F. In some locations, where there is no layer of solid rock over the porous rock, the heat from the magma cannot accumulate. It rapidly dissipates through the upper porous stone to the surface and is transmitted to the air. In these places the lower porous rock does not rise

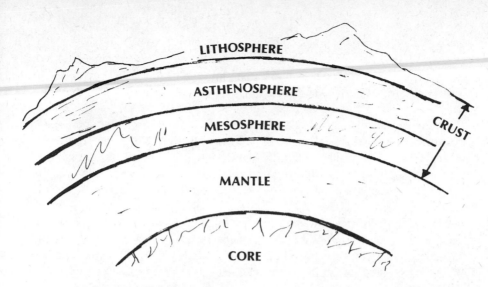

FIGURE 58. The crust of the earth is divided into various layers, each of which has a unique physical makeup.

FIGURE 59. Three forms of energy (steam, hot water, hot rock) are products of the actions occurring in the crustal portion of the earth.

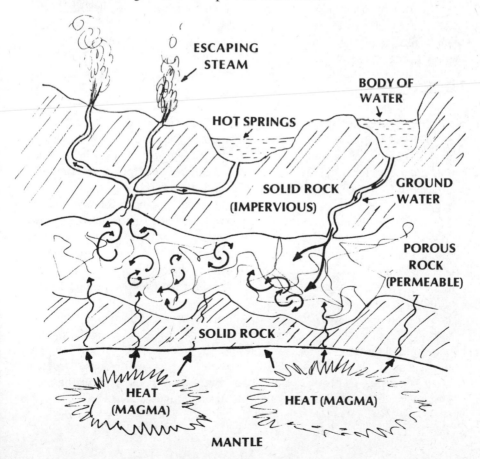

to a significant temperature. The upper layer of solid rock acts like the cover on a pot. It allows the heat below to accumulate and the temperature to rise. Now we introduce a new element—water.

Ground surface water, such as streams and lakes, often percolates down through cracks or fissures in the upper level of solid rock, sometimes to depths of several miles. When the water reaches the level of porous rock, it begins to spread out in all directions. Eventually, it becomes as hot as the surrounding rock, but because of the great pressure at that depth, the water, although far above the boiling point (at surface pressure), remains a superheated liquid. However, just as there are fissures that act as entrances for water in the solid rock, so there are other fissures that allow the superheated water to escape. The hot water, under great pressure, races up the winding passageways to the surface. But as it does, the pressure begins to drop and the superhot water changes or flashes into almost pure steam. Still under a great deal of pressure as it reaches the surface, the hot steam leaps high into the air, creating a fearsome column, such as the tall billowing columns first seen by William Bell Elliott more than a century ago.

Not all geothermal energy exists in the form of pure steam. If the porous rock below does not reach a very high temperature, the hot water ascending to the surface may not contain enough energy to flash it completely into steam. In some cases it may emerge from below as very wet steam—that is, containing a small amount of steam and a lot of hot water. It can also reach the surface as purely hot water (the hot springs in Figure 59). This water may be in the neighborhood of 125° to 150° F. and contain a significant amount (up to 30 percent) of sulphur and other minerals. Such hot springs are common in Yellowstone National Park.

The rock level at which water is heated to form steam need not be of the porous type. In certain parts of the world, because of seismic or other geological activity, large layers of solid, impervious rock are fractured (broken into relatively small pieces) and therefore provide the same permeability for water as porous rock. In other instances there is porous rock all the way to the surface, which should dissipate the heat from below, but over thousands of years the rising hot water has carried with it large amounts of minerals and deposited them in the upper portion of the porous rock, causing the rock to become clogged or sealed. The upper level has therefore become impervious and allows the normal process to occur below —that is, the heating of water and producing of steam.

Geothermal energy, therefore, falls into three basic categories: (1) vapor-dominated sources (pure steam); (2) liquid-dominated sources (containing large amounts of water); and (3) hot dry rock (where water is not present or has not been able to percolate to the heated rocks). In each case it is possible to produce electrical power or supply energy in the form of heat for other uses. Let us take a look at the individual systems.

VAPOR-DOMINATED SYSTEMS

The ideal and by far the easiest and most economical form of geo-thermal energy is pure steam. Unfortunately, this type of geothermal energy is rare; few places have been discovered thus far that supply pure steam. Present estimates indicate that only 5 percent of all geothermal sources consist of dry steam. Luckily for the United States, one of these is located in the Mayacmas–Clear Lake region north of San Francisco called the Geysers—here the geothermal well supplies 99 percent pure steam. Another area that contains pure steam is in the Larderello region of Italy; a third one is in Japan. Figure 60 illustrates a simplified version of a vapor-dominated geothermal power system. Because of the intense heat of the porous rock 1000 feet below the surface, superheated water enters a drilled well at temperatures of 350° to 550° F. and at pressures of 500 pounds per

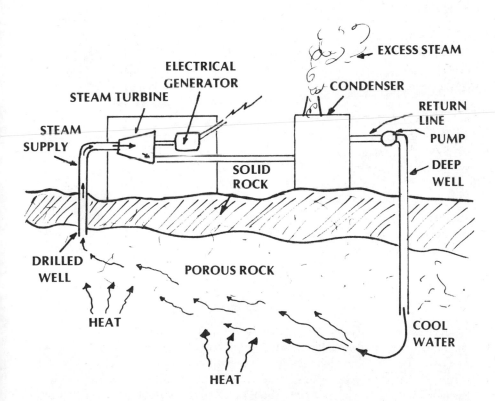

FIGURE 60. The vapor-dominated system (pure steam) is the simplest of geo-thermal power systems.

square inch (PSI) or more. As it rises to the surface, under normal atmospheric pressure, the superhot water flashes violently into scalding steam. This steam, which has now dropped to a pressure of about 100 PSI and a temperature close to 350° F., is used to rotate a specially designed low-pressure turbine (fossil-fueled steam turbines operated at pressures of 2000 PSI and higher). The turbine is connected to an electrical generator, which produces electricity. After it passes through the turbine, the steam is passed through a cooling tower or condenser. Much of the clean steam is vented to the atmosphere. But the portion that contains harmful chemicals (such as boron, ammonia and arsenic) is condensed back to water and reinjected, via a high-pressure pump, into the ground by way of a deep well.

In actual operation steam coming from the ground cannot be fed directly to the turbine. Although it is fairly clean, it still contains small amounts of abrasive material that could, after extended periods of operation, damage the steam turbine. Therefore the steam is passed through a cleaning operation before being passed through the turbine. This was the big problem in 1922 when drillers tapped the steam but did not know how to avoid the turbine damage caused by the abrasive material in the steam. But in the mid-1950s two small companies, Magma Power and Thermal Power, teamed up to drill for steam power on land they had leased in the Geysers area. It was not long before they were actually selling steam from the earth to an electrical utility—Pacific Gas and Electric. By 1960 PG&E was generating 11 MW of electricity from steam generated by the earth's natural heat. Each one of the company's many generating units is being fed by about ten steam wells, some of which go down 8000 feet, spread over an area of 8 square miles. In 1965 Magma-Thermal joined forces with Union Oil of California in order to conduct further exploration and to expand the power-producing capabilities at the Geysers. But since vapor-dominated sources are limited, many prospectors and engineers are looking to the second type of geothermal power, which can supply a great deal more energy because it is far more plentiful.

LIQUID-DOMINATED SYSTEMS

There are probably 20 times more sources of liquid-dominated geothermal energy than are dry steam sources. In this type of resource a great deal of water is located in conjunction with the source of heat. The net result is that large quantities of water accompany the steam as it rises through fissures to the surface. Such a system makes it more difficult to extract the steam necessary to operate a turbine, but liquid-dominated systems still can be of great value. According to Dr. T. David Riney, manager of geophysics and materials for the S-Cubed (Systems, Science and Software) Company, "Certain areas of the United States have vast reservoirs of hot water which can be raised to the surface and converted into electrical power. This technique could be a major step in assuring a plentiful and economical source of electricity, no matter what happens to the world oil supplies."

The reservoirs referred to are large pockets of superheated water with temperatures ranging from 350° to 700° F. When it reaches the surface, this combination of steam and water usually separates by flashing into about 20 percent steam and 80 percent hot water. Figure 61 shows a simplified version of a liquid-dominated geothermal power system. It is fairly similar to a vapor-dominated system, except that the steam-water entering the system must be sent into a special assembly called a separator. In this unit the steam and water are completely separated. The water is either emptied into a nearby river or lake (as in the case of the Wairakei power station in New Zealand) or pumped back into the ground to recharge the original source of geothermal energy. The steam that leaves the separator is sent to the steam turbine to operate the electrical generator. Most new approaches to liquid-dominated goethermal systems are based on the assumption that the water fraction from the energy source will either

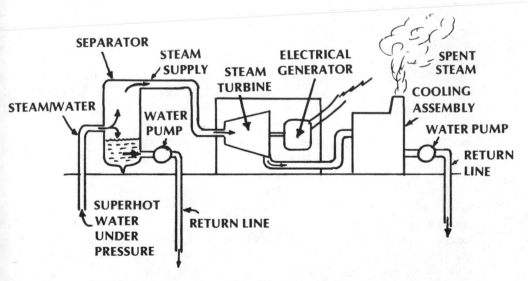

FIGURE 61. Liquid-dominated geothermal sources are more common than dry steam. But these power systems are more complex and less efficient.

be reinjected into the ground, to prevent pollution of local water, or be processed in order to extract the valuable minerals it contains. After processing, the clean water can be added to local water supplies. In some systems the water-steam mixture is double-flashed (separated a second time) and the additional steam is used to operate a lower-pressure turbine —a more efficient method since it extracts more heat (in the form of steam) from the combination. When double flashing is used, the resulting water is usually much cooler (about 212° F.) and represents about 70 percent of the water that came out of the ground (as opposed to 80 percent with single flashing). This means that more steam was produced to drive the turbines. The resulting water can be discarded in a number of ways. In the New Zealand operation the water has a salinity (salt content) only 10 percent of seawater, so it is emptied into the nearby Waikato River with negligible environmental effects. But in El Salvador, where the salinity of the water is 50 percent that of seawater and therefore cannot be dumped into the local river, the waste water is injected (at about 320° F.) into the ground. The possibility of carrying the water by channel to the open sea, some 20 miles away, is being investigated here.

The first power plant to use wet steam was built in New Zealand in the early 1950s. Since this country is short on fossil fuels, it turned to other natural resources, hydropower and geothermal power, at its huge geyser area at Wairakei. The success in New Zealand was followed by a project in Baja California (Mexico) in the southern extension of the Imperial Valley. There, in the Cerro Pietro water steam fields, an electrical power plant has been constructed which will eventually produce 75 MW of power. Attempts have been made to use a wet-steam reservoir in the northern (U.S.) part of the Imperial Valley. However, the hot water contains so much dissolved mineral matter—as much as 25 percent by volume—that it corrodes pipes, clogs turbines and even the wells themselves. But these are problems that can be solved with a reasonable amount of research.

Potentially, a great deal of electricity could be generated by using the energy from liquid-dominated geothermal reservoirs. But many locations are considered impractical because of their relatively low temperature and the very high level of corrosive minerals in their water. The basic problem stems from the fact that the minerals mix with the hot water (the mixture is called brine) and flow, in suspension, up the well (usually an 8- to 10-inch pipe). As the mixture reaches the low-pressure area near the surface, it suddenly flashes into steam. When this happens, the minerals that were in the brine are separated from the water and deposited on the walls of the well (or on the turbine if that is where the flashing occurs). After a while the continuous deposits reduce the diameter of the well, eventually clogging it, and the well must be redrilled, an expensive procedure. The key to the whole problem is the flashing of the brine; if this could be prevented, much of the problem would be solved. But how? To prevent the brine from flashing, it would have to be kept under constant pressure. How can this be accomplished? One way would be to place a pump at the bottom of the well and pump up the brine at high pressure. But the brine would still have to flash before it entered the turbine (in order to create steam to drive the turbine). So a closed-loop system was devised which would transfer the heat to a secondary system (see Figure 62). This is also called a two-cycle or binary system. The hot brine rising from the well is piped into a unit called a heat exchanger. This unit allows the heat from the brine to pass to a liquid flowing through a series of coiled pipes within the heat exchanger, which is the interfacing device between the primary system and the second-

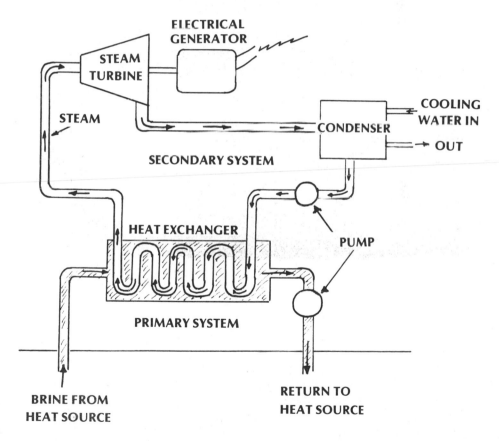

FIGURE 62. A two-cycle (binary) geothermal system isolates corrosive brine from the steam turbine.

ary system. The liquid that is flowing through the secondary system is converted into steam and piped to the steam turbine. After that, it is passed through a condenser, where it is once again cooled down to its liquid state, then returned to the heat exchanger to pick up more heat, and so the cycle is repeated. The liquid in the secondary system can be either water or a chemical such as freon or isobutane.

Such a system is fine; the hot corrosive brine never gets near the steam turbine. But one important element is missing—a method for keeping the brine under high pressure throughout its flow through the primary system. The idea of placing a pump at the bottom of the well is good, if one could find a pump that can survive the extremely destructive conditions in the geothermal reservoir. The pump would have to survive for years in a temperature of 700° F., surrounded by extremely corrosive brine under high pressure. No electrically operated pump could last very long under these conditions. The amount of energy needed to operate a hydraulically driven pump would make the entire system highly inefficient. And it is absolutely ridiculous to consider the use of a long mechanical shaft to drive the pump, which might be as much as 8000 feet below the surface. The problem offers quite a challenge.

But for Hugh B. Mathews, a scientist at the Research Center of the Sperry Rand Corporation, there had to be an answer. After a mental struggle with the baffling problem, Mathews got a brilliant idea. Why fight the system? Why not work with it? Let the energy from the geothermal system do the work of driving the pump. So he sat down and invented the ingenious geothermal pump illustrated in Figure 63. A small turbine is located inside a metal case; behind the turbine is a hollow steam chamber; connected to the turbine shaft is a small pump assembly. The entire assembly is mounted on the end of a steel pipe and lowered into the larger pipe of a geothermal well. Located within the smaller steel pipe is still another small steel pipe, which winds around the middle-sized steel pipe a dozen or so times and finally connects to the hollow steam chamber situated behind the small steam turbine. Here is how it works. Fresh water is pumped down the smallest pipe. As it gets near the bottom, it reaches the section where the small pipe coils around the middle-sized pipe. But this part of the assembly is right in the middle of the superhot brine. The heat changes the fresh water into superhot steam. The steam, now under high pressure, enters the chamber just behind the small turbine wheel. The only way for the steam to escape from this chamber is through the turbine. So it blasts its way past the turbine, causing the turbine to spin at a high speed. Now we have the steam turbine spinning; but the pump assembly is connected to the shaft of the turbine, so the pump is also rotating at high speed. Since the pump is submerged in the brine, it sucks up the brine and

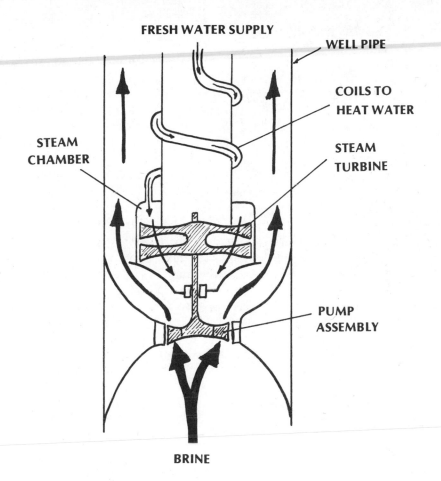

FRESH WATER SUPPLY

WELL PIPE

COILS TO
HEAT WATER

STEAM
CHAMBER

STEAM
TURBINE

PUMP
ASSEMBLY

BRINE

FIGURE 63. A simplified diagram of the Mathews pump (invented by Hugh B. Mathews of Sperry Rand Corporation).

pumps it, under high pressure, up the large well pipe. It can pump up to 1500 gallons of brine in one minute. As long as fresh water is fed to the coils, steam is generated. As long as the turbine rotates, the pump is driven. So the system operation runs on two simple elements—fresh water and the heat of the geothermal reservoir—a brilliantly simple solution to a frustrating problem. With the primary system under constant high pressure, the brine never approaches the flashing point so the minerals in suspension are carried back down to the reservoir after they pass through the primary system. Now that the problem of mineral deposits and clogging of the well pipe is basically solved, a lot more geothermal sources can be considered as candidates for electrical power generation.

Within the category of liquid-dominated geothermal energy falls a relatively new type of reservoir called the geopressurized brine. Deep within the crustal zone of the earth there are regions where large pockets of water have been trapped in highly porous stone for millions of years. This water is continuously heated but cannot escape because it is surrounded and sealed in by impermeable rock. So it forms a type of natural pressure cooker. Because of the depth of these pockets (two to five miles down), and because they are totally sealed off, the pressure builds up to abnormal levels (10,000 PSI) and the temperature of the water reaches 550° F.

Geopressurized brine reservoirs have been encountered by drillers exploring for oil along the Gulf of Mexico. In fact, there may be a series of the huge pockets of high-pressure, high-temperature reservoirs along the Gulf Coast, enough energy to produce a significant amount of electricity. But, unfortunately, we know very little about these natural formations. They are difficult to get to because of their depth, and we do not know exactly how much energy each reservoir contains or how long one would last. A good deal of research will be necessary to answer these questions. But one thing is certain: we are beginning to find out that this planet has a lot of sources of natural energy. If we approach the subject of energy from this point of view, we will be able to find our way out of our present energy and pollution crises.

Geothermal systems operate at very low efficiencies (about 10 percent) because of their relatively low temperature. Furthermore, geothermal reservoirs with water at less than 325° F. are not considered capable of producing electrical power; in operating geothermal power plants, water is discarded (or cooled) at about 300° F. But there is no reason why water as low as 200° F. (or even lower) cannot be used to generate electricity. It is a matter of value (i.e., the value of heat energy in any form) and of system design. As we shall see in the next chapter, a fluid raised merely 10° F. in temperature can be used to rotate a specially designed turbine. Certainly, then, the raising of a fluid 50° to 200° F. can result in a meaningful amount of power generation.

A number of scientists concur on this point and research and development are proceeding in this area. Systems have been designed which can handle low-temperature liquids (under 200° F.). The binary cycle shown in Figure 62 does exactly this. The low-temperature fluid is passed through a large tanklike chamber in which are located a number of coils of tubing from the secondary system. Within these coils is a fluid such as freon or isobutane, whose boiling point is lower than that of water. In other words, these fluids will expand very rapidly with the addition of relatively small amounts of heat. Within the heat exchanger, the heat from the brine that surrounds the coils is transferred through the coils to the secondary-system working fluid. The fluid is rapidly expanded into steam, which is used to drive a turbine. There are a number of additional advantages to this system besides its use of low-temperature brines. Power turbines designed for isobutane or other similar fluids are smaller in size and have fewer turbine stages than do conventional steam turbines because isobutane in vapor form has a higher density than water steam. Furthermore, since the turbine in a binary-cycle system never comes into contact with the highly corrosive brine, it does not have to be constructed of expensive corrosion-resistant materials. Because isobutane is flammable, special seals are required, but this poses no serious problems.

Low-temperature geothermal reservoirs exist in a large number of areas around the world, and these sources of energy should be used. This type of system is already operating at Kamchatka, U.S.S.R., where a power plant using freon as a working fluid extracts energy for producing electricity from water at a temperature of only 180° F. A similar system is working in Japan. In the United States, Magma Incorporated has developed a prototype low-pressure turbine for a 25-MW electrical generating plant to be installed at California's Salton Sea region by the San Diego Gas and Electric Company. One of the advantages of the Magmamax system, as it is called, is that the waste water resulting from the production of electricity will be only 85° F., an excellent solution to the problem of heat pollution. Incidentally, this type of system—when fully developed—will be a great boom to the entire electrical generating industry. It will mean that all the waste steam and hot water from fossil-fuel and atomic plants can be reused to generate additional electricity. Low-temperature systems (which we earlier referred to as bottoming cycles) will make conventional power plants much more efficient in their use of heat energy.

So much for liquid-dominated reservoirs. Next we come to the area that could put geothermal power in the big league.

DRY HOT ROCK

Previously we stated that there was 20 times more geothermal energy available in the form of hot water than in dry steam, which makes hot water a large source of energy. There is ten times more geothermal heat available in the form of hot dry rock than in dry steam and hot water combined. But unlike the other two forms of energy, dry rock systems do not give up their heat very easily. They do not deliver it up to the surface; one has to go down below the earth and harvest it. Heat from various sources in the earth raises the temperature of large bodies of solid, impermeable rock to between 350° and 1400° F. In this way huge amounts of heat are accumulated by the rock. However, since this all takes place in the absence of water, neither steam nor hot water is produced, so it is not possible to drill a well to allow the energy to escape. Nevertheless, there is a tremendous amount of energy—but how to get it? One method has been recommended by the scientists at the government's Los Alamos Scientific Laboratory. It is their idea to drill a hole down to the hot rock level and explode a nuclear device that will fracture (extensively crack) the rock. (Another method of cracking the hot rock is by high-pressure hydraulic fracturing, which has been used for years in the oil industry.) After being fractured, the hot rock is porous to the extent that water can flow through its cracks and fissures. All that is missing now is water. So a well is drilled down to the fractured hot rock level (see Figure 64). A steel pipe with holes perforated near the bottom is fitted down the well. Water is then pumped under pressure (about 7000 PSI) down this pipe and allowed to flow through the fractured hot rock. The water will quickly reach the temperature of the rock and form superhot water. A second well is drilled about 20 or 30 feet from the first and a pipe is fitted into this well. The steam and hot water generated around the hot rock flows up the second steel pipe and into a power plant for use in operating a steam turbine.

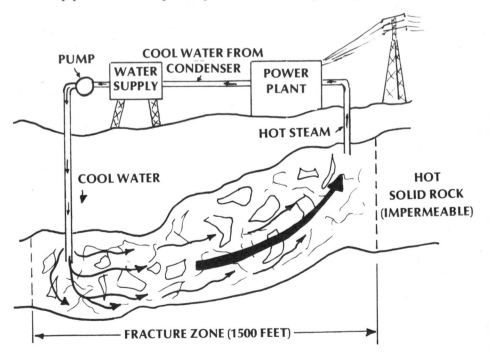

FIGURE 64. Hot dry rock systems can supply ten times more energy than can dry steam and hot water combined.

According to the scientists, once a moderate temperature difference has been established between the cold water descending in the first pipe and the hot water rising in the second, circulation through the system should continue automatically by natural convection. The pump could be turned off except when additional water needs to be injected into the well. In addition to the water being pumped down to the dry rock, additional material such as sand or high-strength glass beads (called propping agents) could be injected into the hot rocks. This material would become wedged into the cracks and fissures, preventing them from closing up again. The size of the fracture zone would be about 1500 feet, but eventually, as the heat is removed from the hot rocks, they will tend to shrink. This would result in new cracks, extending the size of the original fracture zone. Thus a well may be capable of producing geothermal heat for 30 years or more. In fact, under the right operating conditions, a well may provide heat energy indefinitely. Of course, this entire approach to geothermal power assumes that the granite will be impermeable enough to prevent the water under pressure from draining away. It also assumes that the hydrofracturing method works on very hard granite. But these are not outrageous assumptions considering today's technology; it is just a matter of testing. And it is worth the effort since so much energy is available in this form. The first necessity is to locate and explore the various geothermal deposits around the country.

GEOTHERMAL EXPLORATION

As evidenced by surface indications such as hot springs, geysers, volcanoes, and fumaroles (escaping natural steam), geothermal energy reservoirs lying within 20,000 feet of the earth's surface are located in many parts of the world, and a large number of them are in the United States. In the United States there are two distinct dry steam sources thus far identified and developed at the Geysers in California: a small reservoir at a depth of 1500 feet and a large reservoir that covers an area of over 10 square miles. The small well has been supplying about 100,000 pounds of steam per hour, while the large one provides more than 350,000 pounds per hour. And this is only the beginning. Although most of the readily accessible geothermal reservoirs are in the West, according to Dr. Morton C. Smith, head of the American Nuclear Society, there are a great many hot dry rock geothermal energy resources all over the country—even on the East Coast. He cites geology surveys that indicate that geothermal reservoirs, including 340° F. water, exist at depths of 20,000 feet in Pennsylvania and in upstate New York, a boon for the energy-demanding Northeast. Currently, ERDA is evaluating Idaho's Snake River Basin and the Raft River Valley in south-central Idaho as sites for a geothermal energy pilot power plant. Similar studies are being conducted in northeastern Nevada. All these areas that have exhibited geothermal potential are referred to as KGRA (known geothermal resource areas). A huge KGRA is believed to exist in Marysville, Montana, another lies beneath a dormant volcano near Los Alamos, New Mexico, and there is still another in Oregon. But there has not been much large-scale geothermal exploration. According to the U.S. Senate Subcommittee on Water and Power Resources (in a special report, U.S. Government Printing Office, No. 22–211), the government has not been providing any leadership in the development of geothermal energy. Because there is no agency to give unified direction, research and development are sporadic and uncoordinated.

If geothermal resources can supply so much clean energy, why have we not used them to a greater extent? For one thing, until recent years it was believed that geothermal reservoirs were strictly localized and that hot springs and geysers were anomalies—freaks of nature. If this was so, geothermal heat could never supply a significant amount of energy to the United States. But, as we shall shortly see, this is definitely not so; we have in the United States enough geothermal energy to account for a good percentage of our power needs. Another reason for not developing this energy source is that fossil fuels were plentiful and inexpensive. That situation has long since vanished. A third reason stems from the fact that we know very little about geothermal power. We have little experience with geothermal power plants, we do not understand the dynamics of geothermal reservoirs and we know practically nothing about geothermal exploration. But the increasing demand for energy coupled with the depletion of fossil fuel supplies now makes geothermal energy an area with great appeal and promise. Leading the way in the geothermal rush is a new breed of prospector.

How does one go about detecting a geothermal reservoir several thousand feet below the surface of the earth? Early prospectors depended on surface conditions for clues. The world's first geothermal wells were drilled in Italy in the late 1820s. A century later prospectors were drilling in California. In both cases they drilled where they found hot springs and geysers. Obviously, we need a better method than this to find new sources. The new techniques involve such disciplines as geology, geochemistry and geophysics. Geothermal prospecting is a young field untested by the experience of successes and failures. But we have already begun to learn some basic facts about geothermal energy that give us some good clues as to its location.

For example, all the shallow geothermal reservoirs we have located to date are near the margins of special areas known as plate boundaries. The term *plate* refers to that huge portion of the earth's outer crust, the lithosphere that virtually floats or rests on the inner portion (asthenosphere). The entire outer crust of the earth is made up of a number of these plates. Some of them, such as the Pacific Plate and the South American Plate, are huge, covering millions of square miles of the earth's surface. In theory, the earth's continents once comprised a single land mass—the ancient continent of Pangaea. South America butted against Africa, to which were attached Antartica and Australia. Then, about 200 million years ago, Pangaea started to break up. Large sections of land (plates) began to drift away from each other. These plates have never stopped moving, and it is their movement that causes volcanos and earthquakes. The inner portions of these plates are largely free of earthquakes and other seismic activity, but at the edges or boundaries three different motions can occur. Plates can pull away from each other, butt into each other or move sideways and scrape against each other (as in the case of California's San Andreas Fault). And geothermal reservoirs are found along these boundaries. In boundary areas (where two plates meet) magma (molten rock) is generated by the tremendous friction of the plates as they drift past each other. A great deal of study is needed to understand the nature of plate activity (plate tectonics) and the activity that occurs at the boundary areas. The nature of a geothermal system is very dependent on the physical characteristics of that area—porosity, permeability, heat flow, etc. Then one must be concerned with the way water acts in such a system. Hot water rises buoyantly along the path of least resistance. This path of flow may not be straight up; it may travel sideways for a long distance before emerging at the surface. Sometimes it travels for a few miles in a horizontal direction, which makes geothermal exploration difficult because surface indications may not, in fact, reveal the actual location of the geothermal source.

The objective of geothermal exploration is to locate large areas of relatively shallow hot rock, estimate the volume, determine its temperature and determine the condition (permeability) of the overlying rock so that a prediction may be made concerning how deep a well should be drilled to tap a dry steam source or a mixture of hot water and steam. Furthermore, the chemical makeup of the fluids must be determined. To accomplish all this a slew of electronic devices and other detecting methods are employed, including seismic reflection methods, electromagnetic techniques, geomagnetic-variation soundings and heat-flow surveys. All this highly complex activity has one goal: to understand the origin, migration and the major concentrations of geothermal reservoirs as well as we understand the same aspects of petroleum resources. We still have a long way to go. As of

1975 we have drilled approximately 135 geothermal wells (according to James B. Combs, associate professor of geophysics at the University of Texas), whereas we have drilled over 141,000 oil and gas wells.

One way to find geothermal deposits is to use special electronic equipment capable of hearing the sounds of small earthquakes called microearthquakes. These miniquakes often occur below the earth in plate boundary areas—and this is where geothermal sources are usually located. Seismic noise detection techniques can be a major aid in helping to locate our large geothermal resources.

Another useful tool in geothermal exploration is geomagnetic-variation sounding. In this method electronic equipment is used to measure the magnetic field below the earth. Changes in the strength and polarity of the field can give important clues as to the nature of the area being measured. Heat-flow measurement is another handy tool. Very sensitive instruments can detect the slightest variations in the heat that is rising from the ground, which may tell us whether or not a geothermal reservoir is present.

These scientific methods take a lot of the guessing out of exploration. But more work has to be done to perfect these methods in specific application to geothermal energy. According to Professor Combs, "Geothermal exploration is in fact still in the most primitive stage. We are in a stage similar to that of the petroleum industry in the 1930s and 1940, when oil prospectors were drilling oil seeps—that is, we are doing little more today than drilling hot springs."

The United States Congress enacted Public Law 91–581, "The Geothermal Act of 1970," which provides for the lease of public lands for the development and utilization of geothermal steam and associated geothermal resources. In this way the government hoped to stimulate wide exploration of America's geothermal deposits. One company, American Geothermal, has been exploring on 60,000 acres of privately leased land, and has applied for 20,000 acres of public land. It has identified a number of very promising sites. But it costs up to $150,000 to develop one site because geothermal drilling tends to be more expensive than drilling for oil wells, since it is in areas of hard rock rather than in conventional sedimentary basins. American Geothermal obtained extra cash by going into partnership with the Chevron Oil Company (a subsidiary of Standard Oil of California).

If it is so expensive and so little is known about geothermal power, why go into it? Mainly because geothermal power is so promising. There are at least 1000 known hot springs or steam vents in the western states. According to Bernardo Grossling of the U.S. Geological Survey, there are as many as 100 geothermal reservoirs in the United States capable of producing 500 MW of electrical power—and probably a lot more reservoirs will be discovered as we improve our exploration techniques. That is why oil companies and private exploration outfits are sending out more prospecting teams every day. Wildcat steam wells are popping up all over the West.

There are indications that although we have located some very large deposits, such as the Imperial Valley (with its 2 million or more acre-feet of hot water), we may not yet have found the best deposits. Take the experience of David Blackwell, a young geologist from Southern Methodist University. He was studying heat flow patterns in the Pacific Northwest in 1969. He decided to take some readings in an abandoned gold-mining camp, about 20 miles northwest of Helena, Montana. Although there were no telltale ground indications of a geothermal deposit, Blackwell recorded the highest gradient (profile or change in heat flow) ever found in the United States. Normally, the temperature of the earth increases 1° F. per 100 feet of descent. Where Blackwell was testing, the temperature was increasing about 10° F. per 100 feet. So here was an area that had not been classified as a KGRA or even a PGRA (potential geothermal resource area), yet it was hot rock (as high as 800° F.) covering 12 square miles.

Incidents such as this have led the Interior Department's Bernardo Grossling, who wrote the geothermal section of the National Petroleum Council's energy study, to suggest that the United States drill a network of exploratory wells (about 3000 wells each 500 feet deep) across the entire country to determine the full extent of our geothermal energy. David Blackwell recommends a less expensive approach: just measure the temperature gradients of all the drill holes (old gas and oil wells) in the United States that are more than 100 feet deep.

There have been other new discoveries. Dry steam wells have been found in Valles Caldera, New Mexico, representing a second United States dry steam field. Other large dry steam deposits may be hidden beneath the surface of the earth in other states. From surface indications alone there appears to be a long belt of geothermal reservoirs stretching down the western coast of the Americas, from Alaska, through California, Mexico, all the way to Chile. Yet, with a few exceptions, America's utilities and other energy companies are not actively participating in this area. Why?

In the past, two industries have committed risk capital to develop our natural resources on the scale presently required for geothermal development—the mining and the oil interests. It would seem natural for the oil companies to follow up on this new energy source, but they have not because while oil can be stored, shipped and sold to all markets in the world, geothermal energy must be used immediately, on site. This makes it an inflexible product. On the other hand, with the depletion of domestic oil supplies, oil companies that do not have Middle East or other world sup-

Geothermal steam field at the geysers, just north of San Francisco. Steam is used to operate the Pacific Gas and Electric Company's generators, which produce 396,000 kilowatts of electricity. (Courtesy PG&E)

plies might be wise to develop geothermal steam for sale. Furthermore, this energy source need not be a site-bound product. As we enter the hydrogen era, electricity generated from geothermal energy could be used to produce hydrogen, which can be shipped and sold in other markets.

The utilities presently do not have the huge sums of capital necessary to develop a total geothermal energy industry. This large burden will have to be carried, initially, by the federal government. And that is beginning to happen, but not fast enough. Although the Energy Research and Development Administration budgeted $55.8 million in fiscal 1976 (up 106 percent over 1975), even Congress agrees a lot more funds will be needed.

Congressional *Geothermal Resources Report 22-211* states, "Lack of information about geothermal resources, coupled with insufficient experience in drilling and production techniques, make geothermal development risky ventures. Those private firms which would like to undertake geothermal projects find financing difficult to obtain. Until more experience has been acquired through the construction and operation of demonstration facilities, private financing for geothermal exploration and development will probably remain difficult."

PROBLEMS AND SOLUTIONS

We are not sure that the technique of fracturing hot rock and expanding the fracture zone will actually work. We also do not know if fracturing will cause seismic activity, especially since geothermal reservoirs are in earthquake-prone areas. These uncertainties will have to be settled.

Subsidence is another problem. In hot-water geothermal power plants huge amounts of water are extracted from the ground. At the Wairakei fields in New Zealand, 70 million tons of water are extracted from the earth each year. As a result, the earth in some areas began to literally cave in or subside. The seriousness of this problem depends on the natural structure of the earth at a given site and the amount of water extracted and the rate of extraction. But the subsidence problem can be solved either by limiting the amount of hot water extracted to a safe rate, or by using a binary system and reinjecting (recharging) the water into the earth. In areas such as California's Imperial Valley, where a geothermal resource is adjacent to an agricultural area with irrigation systems, subsidence must be carefully controlled to prevent the destruction of the irrigation systems.

Another tricky problem in geothermal power is the high mineral content of hot water (and steam/hot water) reservoirs. Dozens of chemicals are present in the water, such as boron, ammonia, arsenic, mercury, silica, carbon dioxide, hydrogen sulfide and salt, and they can cause a number of problems. First, they corrode the entire geothermal system, coating turbines, clogging pumps, and closing up well holes. However, at the Mitsubishi geothermal steam station, there is no major corrosion problem. Once or twice a year the turbine blades are scraped to remove accumulated salts, but this operation is no worse than routine operations associated with other types of power.

Where mineral content is extremely high (some water contains as much as 30 percent minerals) the corrosion problem can practically be eliminated by isolating the corrosive in a binary system; the water from the well will then remain in the primary system and never come into contact with the turbine.

Chemicals in geothermal fluids can also create air and water pollution. Where steam vapors are exhausted directly into the air, the exhaust must be processed to remove all noxious or otherwise harmful pollutants—hydrogen sulfide, for example (a poisonous gas). But with extraction, the end product will be pure sulphur, which is salable. When water from geothermal wells is emptied into local surface water, there is a pollution threat. However, a study carried out by Professor R. C. Axtmann of Princeton University at the Wairakei geothermal power plant showed that this is not necessarily a serious problem. When the waste water mixed with the river, the resulting concentration of pollutants was well below the maximum recommended for potable water.

All air and water pollution problems can be eliminated from geothermal power systems by pumping the fluid back into the ground. Professor Axtmann asserts, "Although optimal extraction and optimal utilization are separate issues, there is a potential technical linkage between the two: reinjection of the hot waste water would decrease the net energy drawoff and thereby increase the thermal efficiency and, possibly, prolong the life of the reservoir. Moreover, many of the most serious environmental shortcomings of the plant would disappear, for example, arsenic, mercury and some thermal contamination." This also applies to all the noxious and poisonous gases that might be vented to the atmosphere. But what will be the effect of reinjecting the waste water into the ground? They have been doing it at the Geysers plant in California for a few years now with no apparent harm.

Geothermal power can suffer from two associated problems—thermal inefficiency and thermal pollution. Because of its relatively low temperature (600° versus 1200° F. for fossil fuel steam plants), geothermal power plants are not efficient. Therefore the water leaving a plant still contains a great deal of heat. This water (or steam) must either be cooled with condensers or in cooling towers, or dumped into nearby bodies of water. In any case, a lot of heat is introduced into the local ecological system. Here again, the problem can be solved by reinjecting the fluid back into the geothermal reservoir, which will improve the system's thermal efficiency. An additional solution would be to add a bottoming cycle, or low-temperature generating system, to the geothermal power plant. Then more of the heat will be used to generate electrical power. When waste water is reinjected into the ground, there is no need for large amounts of cooling water for heat extraction. Thermal pollution is not a great problem for dry steam geothermal power plants either, since they also return coolant water to the ground. Neither do they have the problem of water availability as do other types of thermal plants (power plants that use heat energy), because they can condense the steam and use the resulting water for cooling purposes.

The point is that unlike other types of thermal plants, geothermal plants do not have to compete with other local users of fresh water.

Still to be developed is a technique for very deep drilling. Although modern drilling methods can get us down 30,000 feet, we may have to go down 50,000 feet in order to tap large amounts of dry hot rock geothermal energy. But this is hardly an insurmountable problem, even though new pumping methods may be required to keep such a well operating efficiently because of friction losses.

Noise pollution has been associated with geothermal power. High-pressure steam escaping from the ground is ear-splitting. But dry steam represents only a small fraction of geothermal power, and these steam fields are distant from heavily populated areas (certainly out of hearing distance).

Critics say that since geothermal power is localized, electric utilities must be situated right at the energy source. Except for line losses, this may not be so bad. Aluminum companies and other users of large amounts of electricity might well build plants near geothermal sites. Also, with the development of superconducting lines, and the use of hydrogen as an energy mover, transmission should be feasible. Many geothermal sites will be less than 100 miles from energy-consuming areas, not an excessive distance to transmit power.

Another criticism of geothermal power plants is that they might occupy large tracts of scenic lands. So far, land for geothermal speculation has not been taken from national parks, and even on the land designated for exploration, only a small percentage will actually be occupied by geothermal power plants. Most of it will remain open, undeveloped land, available for farming and recreational uses.

Electric utilities planning to use geothermal energy will have to build plants in small 55-MW modules (this is the most efficient size for geothermal energy). But this is really an advantage. Other types of thermal plants (fossil fuel and atomic power) must be built very large for optimum efficiency. But with geothermal power, a utility company can build only the capacity it needs at the start (which means less up-front capital) and add 55-MW units as necessary.

One question, which may not indicate a problem, nevertheless must be answered before large commitments of money and labor are made to a geothermal site. How long will a geothermal well continue to supply heat energy? Most experts think geothermal reservoirs will last a significant number of years. The Geysers power plant has been supplying electricity for over 15 years now and is still going strong. Larderello has been feeding the surrounding area power since 1904, and provides some basis for evaluating the life span of a geothermal site. It takes about 25 to 30 years of operation to fully pay off the costs of exploration, well development, power plant design, construction and transmission line installation. A carefully managed well should operate for 100 years or more, and if the well's waste water is recycled back into the ground, it might be able to operate indefinitely. Extracting heat from the earth indefinitely, or even for very long periods of time, however, may eventually cause the earth's inner area to cool off, leading to disastrous results. We have no need to worry, says J. Herbert Anderson, a geothermal power consultant. If electric utilities produced power at a rate of 420,000 MW for 41 million years, we would cool the earth by only 1° F. That ends that problem.

Actually, there are no unsolvable or even extremely challenging technical problems with this energy source. Geothermal power plants require no fuel, they could operate indefinitely without air or water pollution and they do not demand tearing up the earth, as in strip mining, or risking people's lives, as in deep mining. They pose no radiation risks. These thoughts are well stated in the 93rd Congress Geothermal Energy Report 22-211. "To understand properly the impact of the production of electric power on the environment, it is necessary to evaluate more than just the power plant, whether it is geothermal, nuclear, or fossil fueled; the entire fuel cycle from mining, processing, transportation, and the disposal of spent waste must be considered. When viewed in this light, the environmental effect of geothermal generation appears to be minor when compared with fossil-fueled or nuclear generation. The environmental impact of geothermal generation is restricted to the generating site, whereas much of the environmental impact of other power generation takes place at several locations (mines, processing plants, disposal sites). . . . Geothermal energy, even with its environmental problems, remains a relatively 'clean' energy source."

HOW MUCH POWER

Estimates of geothermal resources vary greatly, but most experts agree there is a lot of geothermal power to be had. Central America, for example, may have more geothermal energy than it can ever put to use and may eventually become an exporter of power. The Soviet Union estimates that its geothermal resources may be greater in energy content than its combined reserves of coal, gas and oil. The Afar region in Ethiopia may have enough geothermal fields to provide electrical power for all of Africa. Right now at least 18 countries, besides those already using the earth's heat, are actively developing their geothermal resources.

In the Philippines the Union Oil Company in conjunction with the Philippine government is working to develop that country's huge natural energy resource. Present estimates are that geothermal sources could account for 50 percent of Philippine electrical generating needs within ten years and that geothermal energy will cost a quarter or less than fuel oil does today. The first large-scale natural steam-driven power plant will begin producing electricity at Tiwi on the island of Luzon by the middle of 1977. In New Zealand geothermal energy supplies almost 15 percent of the power requirements. At the Geysers, the initial output of electrical power was only 12.5 MW. Presently it is about 400 MW and there are plans to increase output to 908 MW by 1977 and to 1200 MW by 1980. The total energy potential at the Geysers has been estimated as 3000 MW, which is a considerable amount of power, and some experts believe it may be as high as 10,000 MW. We cannot hope that geothermal power will provide a high percentage of our electricity needs in a short time. But then, our energy requirements are enormous; in 1980 the United States will need over 655,000 MW of electricity. Still, geothermal power can play an important role. As Professor Comb says, "The estimates range from less than 1 percent to 100 percent. But that is not the crucial question. I contend that future efforts to develop techniques for geothermal exploration and utilization should not be justified on the basis of the future electrical power needs of the United States that can be provided by geothermal energy. Geothermal energy may never provide more than 5 or 10 percent of the electrical needs of the United States. However, this small percentage could have considerable impact on the American economy. For comparison, only 4 or 5 percent of the electrical energy in the United States has been (in recent years) or will be, produced by hydroelectric installations. But consider the Pacific Northwest without hydropower; almost all of the power produced in Washington, Oregon, Idaho and Northern California comes from hydropower."

When we look at forms of geothermal power other than dry steam, such as hot water and dry hot rock, we arrive at much more substantial figures. The geothermal resources found in California's Imperial Valley, for example, are considerable, perhaps enough to provide 30,000 MW of power, so this field alone could supply 6 to 7 percent of our electrical power needs. In addition, it has been estimated that as much as 7 million acre-feet of fresh water (by desalination) could be furnished by the Imperial Valley geothermal reservoir. In this water-deficient area the additional water (equal to half the flow of the lower Colorado River) would be greatly welcome. Considering that total world production of geothermal power in 1971 was only 790 MW, these figures are impressive. As new areas of the West are opened for exploration, it is expected that this part of the country will supply about 75,000 MW of power, which will be approximately 11 percent of our electricity needs in 1980. And none of these estimates take into account the full potential of hot dry rock energy. If this aspect of geothermal power were fully developed, it would add significantly to the total energy. Reading Geothermal Energy Report 22-211, we begin to understand the huge amount of energy involved: ". . . in the Western States, geothermal resources have the ability to provide a substantial portion of future electrical demands. It has been estimated that as much as 400,000 megawatts of electric generating capacity based upon geothermal energy could be installed in the United States by the year 2000 if aggressive research and development efforts were initiated quickly. "This amount equals the Nation's total present installed generating capacity of all types."

Bolstering this heartening prediction is a geothermal energy estimate by the U.S. Geological Survey: the present "recoverable" geothermal resources in the United States, excluding Alaska, is of the order of 1 billion megawatt-centuries. This means that we can generate 1 billion MW of power (1000 times the U.S. requirements for 1990) continuously for a hundred years. If we add the eastern two-thirds of the United States (presently not economically recoverable), we could generate this phenomenal amount of power for 1000 years.

Other reports state that we could generate 100,000 MW for at least 1000 years, while some scientists such as Dr. Morton C. Smith (American Nuclear Society) state that geothermal power could provide all the electrical power that the United States will require for the foreseeable future. Certainly, many experts think, we could generate up to 30,000 MW of electricity by 1985 using geothermal heat. J. Hilbert Anderson, a geothermal power consultant, believes the world has a huge source of energy in geothermal heat: "If we could convert 13 percent of this heat to power, we could produce 4,160,000 MW continually. This is approximately ten times the world's present average power output."

The indications are that geothermal power, although still a young field, will hold its own and then some when pitted against its competition. However, economic competitiveness will depend greatly on how efficiently and inexpensively the conversion process can be carried out, and on how much it will cost to gain access to the geothermal energy. In this latter regard, advances in cheap, efficient, high-speed drilling technology will play a vital role.

One big advantage enjoyed by geothermal power is the low capital cost of building power plants. The entire steam generating facility, including expensive, complex boilers, controls and fuel handling systems, is eliminated. Of course, geothermal plants have the added costs of maintaining wells; these wells must be cleaned out periodically and sometimes redrilled at a cost of about $150,000. Still, evidence thus far accumulated suggests that geothermal power is one of the cheapest forms of energy available.

For one thing, the plants can be left virtually unattended for 16 out of every 24 hours (this is the procedure followed in Geysers plant). Because of the simplicity of geothermal energy systems used in household heating and air conditioning, operating expenses can be reduced by 90 percent or more. A hotel in Rotorua, New Zealand, reports that it operates a heating and air-conditioning system for only 12 cents per million kilocalories. Using oil, the same system would cost 20 times more to run. As a bonus, the hotel has no pollution problems to worry about.

At the Geysers power plant the economic picture also looks pretty good. Electrical power had been selling at about 5 mills per kilowatt-hour. Electricity based on oil-fired steam was selling for twice as much (10 mills per kilowatt-hour) when oil was selling at $3.50 per barrel. So-called "cheap" atomic power was selling at 12 mills per kilowatt-hour in northern California. Today, as labor, materials and fuels soar in price, geothermal electricity will stay pretty close to 10 mills per kilowatt hour, which certainly is competitive. (This comparison does not take into account the energy cost to society due to the damage caused by pollution and other aspects of conventional electricity generation activities.) The only method of generating electricity that is cheaper is hydropower, and that is a power source we should not mind having around either.

Now that many scientists and companies realize that there is a lot of geothermal power available and that it is cheap, clean energy, things are starting to happen. Shell Oil has bought two tracts of KGRA (known geothermal resource areas) with a winning bid of $4.5 million. Over 30 companies have acquired government land and more than a dozen of them are already drilling. The San Diego Gas and Electric Company is developing a geothermal steam capability in the Imperial Valley. It hopes to be generating electricity by 1980, which would make it the nation's second utility to do so. In Japan the Mitsubishi Company has a 10-MW geothermal plant that provides electricity for its Akita zinc plant. More and more companies are beginning to allot budgets for geothermal energy development. Besides exploration and well development, most of the funds are being spent on research, including methods for handling hot, corrosive fluids under pressure, new ways to economically separate steam and hot water, effective ways of disposing of waste fluids and testing of new materials. The federal government, in contrast, has been slow to provide financial support for geothermal power. Until recently, its support has been microscopic compared to the amount of money it has spent on nuclear power development—it took about seven years to pass the Geothermal Steam Act. Federal participation in geothermal development has begun to improve, however, and though still not at the level it should be, nevertheless is significant. In 1968 a miniscule $20,000 was the government's contribution to geothermal research and development; in 1972 it crept up to $200,000, which was barely enough to generate proposals for geothermal projects. The now-famous Arab boycott of 1973 pushed the appropriation to a useful but inadequate $14 million. The $45 million for geothermal power designated in 1975 has been a big boost, but not nearly enough. A much more intensive effort is needed; the federal budget must be at least $100 million per year if we are to see the rapid development of a significant amount of geothermal-based electricity by 1980. This is not an unreasonable amount considering the hundreds of millions spent annually on nuclear power.

For thousands of years we have been consuming wood, coal, gas and oil to generate heat. Geothermal energy eliminates all the prior steps and provides a direct source of heat, a huge source that will be with us for millions of years. It will also permit us to reserve great quantities of oil and other fossil fuels for use in industries such as plastics and petrochemicals. As stated by the National Science Foundation, "The estimate which has withstood the scrutiny of the largest range of expertise . . . is that the nation's geothermal resources could be supplying 132,000 megawatts in 1985 and 395,000 by 2000." In 1974 geothermal power was supplying only a hundredth of the nation's electricity. By 1985 it could supply about 15 percent. This would save a huge amount of petroleum—about as much as was used by all 90 million cars on the road in 1974.

This natural source of heat, if fully developed, will provide a good source for basic purposes besides the generation of electricity. In Japan Mitsubishi is studying ways to use the excess heat from geothermal resources for local house heating and for food-producing greenhouses. In Iceland volcanic hot springs have been used for generations to heat homes. And not a few isolated homes either; the city of Reykjavik (population 60,000) is heated almost exclusively by geothermal energy. We ourselves have been putting geothermal heat to good use in a few places. In Klamath Falls, Oregon, about 400 residences, some schools and quite a few businesses are heated by the earth's hot-water reservoirs. In Lakeview, Oregon, a large tomato greenhouse is heated this way, and around 200 homes in Boise, Idaho, are kept comfortable on the coldest winter nights with heat from the earth. Melun, a small area on the outskirts of Paris, drilled to a depth of 5850 feet to tap a large geothermal reservoir, which is now heating its homes and supplying hot water in bathrooms and kitchens. The French government has been so impressed with the success of this project that it has allocated $1.5 million for additional geothermal well drilling. The Russians reported that they were able to save the equivalent of 15 million tons of fuel in 1970 by using geothermal power.

Paper companies located near these natural heat sources can use them to warm water for pulp processing. Probably one of the most important secondary uses of geothermal energy will be the desalting of water to provide fresh water where it is in short supply. Because the water is already heated when it is pumped out of the ground, the distillation (heating the water to get rid of the salt) of large amounts from geothermal reservoirs may prove to be an inexpensive process.

Geothermal energy has everything going for it—it is free, relatively clean, abundant and not susceptible to weather or political change. Once we develop effective and economical techniques for drilling to depths of 50,000 feet and more, vast geothermal resources from all over the country will be within our range of operation. But it is important to remember that geothermal power need not be available throughout the nation in order to be a valuable asset. The Pacific Northwest, which has traditionally relied upon hydroelectric power, is approaching the practical limits of that resource. The area has no appreciable reserves of fossil fuels and could benefit greatly from the development of geothermal technologies. Areas that use this source of energy will not draw from our dwindling supplies of gas and oil or adversely influence our balance of payments by importing foreign oil or liquid gas.

The heat of the earth rises from its core and warms the ground upon which we walk. It has been stored, as though in an energy savings account, and the balance of the account numbers in the countless trillions of calories of heat energy. If we draw on merely the interest of these savings, we will never exhaust our account.

9

A Bright Approach to Electricity

The sun has been around for billions of years, but man is just beginning to think about it seriously as a source of energy: it is, in fact, such a spectacular source of power that it staggers the imagination. When we recline on the beach to get a tan, we are lying beneath a huge ball of burning gas that is the size of 1,300,000 earths. It is no surprise that it feels warm on our skin. The temperature of the sun at its center is about 25 million degrees F. The material at the center of the sun is under fantastic pressure, about one trillion pounds per square inch. It has been calculated that a tiny piece of the sun's material, the size of a pinhead, would emit enough heat to kill a man a hundred miles away. At the sun's temperature any known substance in the universe would be vaporized into a special type of gas called plasma. Every second, the sun throws off into space more energy than man has used since civilization began. Only one two-billionths of it reaches the earth, but in three days this tiny fraction of the sun's energy provides as much heat and light as would be available from all our known reserves of coal, oil and gas.

A solar flare can erupt with the force of a billion hydrogen bombs. If the earth were not protected by a strong magnetic field and the atmosphere, a flare's radiation would destroy all life. Concentrated on earth, its heat could melt the polar caps. A single solar flare may release as much energy as the entire world uses in 100,000 years. One of the most violent events identified by NASA's Skylab was a great explosion in the sun's corona called a coronal transient. This mammoth explosion hurled fiery loops of matter weighing hundreds of thousands of tons into space at speed of more than a million miles per hour.

The sun will maintain its level of activity for at least another 5 billion years, after which it will have used up most of the basic fuel (hydrogen) in its core. Its thermonuclear reactions will then expand outward, where unused hydrogen exists. As it does, the tremendous nuclear heat at its core will also move outward, forcing the sun to expand to as much as 60 times its present size, which will cause the sun to cool and its surface to become deep red. At this point it will be a fearsome Red Giant, looming across much of our sky. As it is dying, this huge red ball will boil off our water and air and incinerate any remnants of life on earth. Eventually the Red Giant will shrink to a white (hot) dwarf, no bigger than the earth, but so dense that a sugar-cube-size piece of it will weigh thousands of tons. After another billion years what was once a raging fireball and source of life will cool and dim to a black cinder—and eternal night will descend on the solar system. But this time frame is so far in the future that the sun can be considered for all practical purposes an inexhaustible source of energy, including electricity.

Electricity from the sun is really not such a strange concept. The production of almost all electricity involves the use of a heat source, and the sun is a fantastic source of heat. Nor is solar power a new idea. Over 2000 years ago (214 B.C.) a Roman fleet of warships sailed to the island of Syracuse confident of a successful conquest. To their horror, and as if by magic, their ships burst into flames. The entire fleet was reduced to ashes. On land, the great mathematician Archimedes watched with satisfaction. It has been his invention, the series of sun-reflecting mirrors, that set the Roman ships on fire at a distance of a bow shot (about 200 feet).

In more recent history, men became even more resourceful in using the power of the sun. Just before the American Revolution a Frenchman named Lavoisier designed and built a powerful solar furnace that was capable of melting a number of different metals. In 1876, in Bombay, a steam engine of 2½ horsepower that operated on the sun's energy was constructed. In 1878 a steam engine operating off the sun's energy was exhibited by A. Mouchot at the Paris World Fair. Another steam engine, which was capable of generating 50 hp, was built in Egypt in 1913. The sun was even used to produce fresh water from sea water by distillation in a 50,000-square-foot solar energy plant erected in Chile in 1870.

Today scientists are becoming increasingly interested in harnessing the energy of the sun because it is a virtually unlimited source of energy that will allow us to produce electrical energy with zero pollution. In July

1973 a four-day meeting of 900 eminent scientists from Japan, France, Greece, West Africa, Australia, the United States and many other nations was held at the UNESCO house in Paris. Presentations were made by scientists on solar applications as diverse as heating bathwater, melting metals for industrial purposes, desalinating water from the ocean, pumping well water and drying industrial timber. Dr. Maria Telkes, one of the world's foremost solar energy scientists, puts it this way: "Sunlight will be used as a source of energy sooner or later. Why wait? Solar energy is available everywhere. It does not need transportation facilities and cannot be cut off by sudden emergencies created by man. Solar energy cannot produce soot, fire hazards or mining accidents. It is the cleanest and healthiest fuel."

At first slow to react, the United States government is now leading the way. The initiative for new legislation has come from members of the Congress, especially Representative Mike McCormack (D-Wash.), chairman of the Energy Subcommittee of the Committee on Science and Astronautics. A former chemist at 'the AEC's Hanford Reservation, McCormack has actively supported solar energy legislation. Farsighted members of Congress, including Senator Hubert H. Humphrey of Minnesota, had introduced solar energy bills back in the late 1950s and early 1960s. But no one was interested at that time and the bills died of neglect.

In early 1974 attitudes began to change rapidly. By September 26 different solar bills had been introduced. Senator Humphrey made an observation on the international ramifications of our developing solar technology: "If we put ourselves to the task, in a short time we can develop solar energy that will not only meet much of our own needs but will be an exportable product. We will therefore be able to help other people who are desperately in need of energy resources and help them in a way that in no way threatens the peace of the world."

Easily spanning the 93 million miles to earth in eight minutes, solar energy floods the daylight side of our planet with an amount of energy—over 1000 watts per square meter, or 173 quadrillion (10^{15}) watts of power every 24 hours—equal to almost 100,000 times the electrical generating capacity of the entire earth. According to Dr. Telkes, "If all the readily available supply of uranium in the earth's crust were to be converted for global energy needs, it would be equivalent to the solar energy received in a tropical land region during three months." In only 15 minutes the illuminated side of the earth receives the same amount of energy as is consumed by all the people on earth for an entire year. Of course, not all the sunshine reaches the earth's surface—some is blocked or absorbed by the clouds and the atmosphere, and some is reflected back into space—but about half of the sunlight does reach the surface.

The sun is fundamental to our very existence, and it should be man's major resource for the energy he consumes. One of the reasons it is not is that the sunshine that reaches the earth is very diffuse. It is spread out thinly over large areas, so collecting enough solar energy to put it to work becomes a problem. Second, because the earth rotates, creating night and day, and because of cloud conditions, solar energy is intermittent and not continuously available around the clock in any single location. Third, solar power remains a very expensive form of energy, as much as five times more costly than conventional energy. The goal is to develop mass production methods and bring down the cost of solar power so that it will be competitive with other forms of energy. With modern technology, this is possible. We have the solutions; all we need do is apply them.

Solar energy can be approached in several ways. For simplicity, we will divide solar activity into three basic areas: domestic energy systems (space and water heating); total energy systems for commercial and industrial buildings; and large-scale solar electrical power generating systems. There are also three basic methods of converting the sun's energy: direct heat, photovoltaic conversion and bioconversion.

People in the warmer latitudes (in Israel, Japan and Florida) have been using rooftop solar heaters to provide hot water for years. Solar energy for home space heating is already competitive with electric resistance heating. When fossil fuel costs double, as expected, solar heating systems will be economically attractive in all parts of the country. Our society will probably begin its use of solar energy with heating and cooling, the simplest systems to start with. (Incidentally, heating and cooling presently consume 26 percent of the nation's energy resources.) Soon after the full development and use of these systems will come the introduction of much larger and more sophisticated solar electrical power systems.

SATELLITE SOLAR POWER

Probably the most dramatic and adventuresome approach to solar power is the earth satellite power station. The idea is to construct a huge satellite in a synchronous orbit (one in which the satellite remains in a fixed position over one location on earth) 22,300 miles above the earth. This is the perfect place for a solar power plant to be since the space above the earth's atmosphere gets the most concentrated amount of sunlight. In fact, it gets about 28,000 times more than all the energy consumed on earth. If we could capture and convert all the sunshine that reaches our planet during a single hour, we would have enough to supply the entire energy requirements of the United States for nine years.

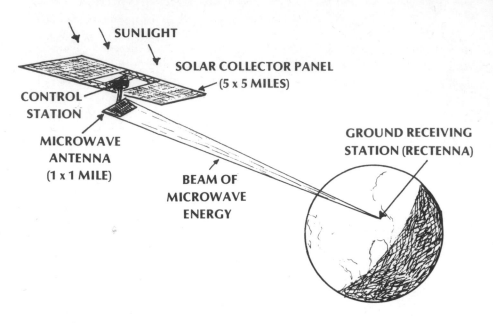

FIGURE 65. Satellite solar power stations (SSPS) would measure 25,000 feet on each side and weigh about 25 million pounds.

The idea for such a satellite was proposed by Dr. Peter E. Glaser, a scientist and vice president of the Arthur D. Little Company. He fully believes that if we make a concerted effort, by the turn of the century a half-dozen satellites could be providing all the power we need for lighting, heating our homes and offices, chilling our refrigerators and spinning every motor in the country. In fact, Dr. Glaser goes so far as to claim that with the world's demand for electricity growing so rapidly, the human race must either use solar energy or live on a decaying planet choked with polluted air and water.

Orbiting solar satellites would be more efficient than ground-based systems. For one thing, the sunlight would be available virtually 24 hours a day, without cloud interruptions. For another, the satellite could convert between 6 and 15 times more energy to electricity than could terrestrial systems. The high efficiency of the special microwave method of transmission and transformation into conventional electricity at the receiving station will reduce waste heat to a fraction of that created in standard earth-based heat engine methods (steam plants, nuclear plants, etc.). Such a satellite, or Satellite Solar Power Station (SSPS), as it is called, would be extremely large (see Figure 65) since it would be required to supply literally billions of watts of electricity. The complete assembly would consist of four main sections or subassemblies, plus the receiving station located on earth.

THE STRUCTURE OF SOLAR SATELLITES

The largest section would be the solar energy collector panel, measuring over 25,000 feet on each side. That means that there would be a flat structure orbiting our planet that would cover an area of about 16,000 acres. The purpose of this panel would be to intercept sunlight and convert it to electricity.

A second large panel would be the microwave antenna, a device that can transmit or radiate energy in a single direction. Although not as huge as the collector panel, it would measure one mile on each side, which still makes it a pretty large object to be placed in orbit. The microwave antenna would receive the huge amount of power via a third portion of the SSPS, a two-mile-long cable that would carry the electrical power generated by the collector panel to the microwave antenna. This cable or transmission line would have to be very heavy since it would be carrying an enormous amount of electrical power. To better appreciate how heavy, consider that whereas the average house wiring safely carries about 3000 watts of power, the satellite cable will have to carry as much as 15 billion watts, the total amount of electrical power used by New York City.

Transmitting huge amounts of power from a satellite to earth on an invisible beam is not science fiction fantasy. The method has already been tried by NASA at a distance of one mile and it worked. Dr. Glaser believes we can transmit billions of watts of power from space with 70 percent efficiency. As usual, there are problems to be solved. One is finding the best frequency to use in transmitting the power. At frequencies above 10 gigahertz (10 billion cycles per second), the energy of the power beam can be seriously reduced (absorbed and scattered) by the atmosphere, especially in rainy conditions. So the frequency chosen must be able to penetrate the atmosphere and cloud cover even under the worst conditions. Right now it looks as if this frequency will turn out to be somewhere between 3 and 4 GHz. The efficiency of the antenna is obviously of prime importance. Also, the size of the antenna will determine how large the entire satellite will be. The most economical range appears to be between 3000 and 15,000 MW of power.

Another difficulty is generating huge amounts of electrical power at microwave frequencies. The solar collector will send DC power to the transmitting assembly. The transmitter unit will have to convert this to high frequency. Microwave generators with outputs of 5 kilowatts and an efficiency of 90 percent have been successfully tested and are now commercially available, but the orbiting station will require at least a million times more power. This could be attained by massing a number of smaller units. It would take hundreds of thousands of these microwave generators working together to transmit the power to the ground station, but the generators are quite inexpensive, only about $30 each.

The final component of the system is the control station subassembly. In this part of the satellite would be located all the controls needed to keep the satellite pointed in exactly the correct direction. Controls would also be necessary to monitor the flow of electricity and point the microwave antenna at the right spot on the earth, plus to operate many other functions in the satellite. Present control technology would permit us to aim the solar collector assembly at the sun with an accuracy of 1 degree, and the transmitting antenna at the earth-based receiving station to within $\frac{1}{60}$ of a degree. This means that from 22,300 miles in space the microwave beam carrying the electrical power will hit within 200 feet of the target center. So

FIGURE 66. A huge antenna called a rectenna would receive the electrical power from the satellite.

there would be no problem hitting the ground station, which will cover an area of about 36 square miles. The receiving antenna, or rectenna, as it is called (see Figure 66), is designed to capture the beam of microwave energy and simultaneously change or rectify it back to DC power so that it can be distributed to consumers by the conventional electrical power grids. The rectenna, which consists of thousands of independent receiving elements, does not have to be accurately aimed at the SSPS transmitting antenna for efficient operation. Nor is the efficiency of the rectenna affected by any distortion of the microwave beam as it passes through the earth's atmosphere. In theory, the absorption (ability to accept microwave energy) efficiency of the rectenna is 100 percent. In practice, the rectenna is expected to have an efficiency of between 85 and 90 percent. One point of interest is that because there is virtually no pollution at the rectenna facilities and no cooling systems are required, ground receiving stations will be well suited for inland areas of the country, especially desert regions. The amount of land that would be required would depend on the quantity of power being generated. But if SSPS systems are to supply a meaningful amount of power around the turn of the century, let us say 500,000 MW, there would be a need for about 50 satellites (each one generating 10,000 MW). For this much power, the ground area required for receiving stations would be about 770 square miles, certainly an insignificant fraction of the arid land that will be available in the year 2000.

At this point one might naturally ask how sunshine gets turned into electricity. The entire process of using the sun's energy depends on a twentieth-century invention called the solar cell. The solar cell is a flat panel coated with a special chemical (see Figure 67) that reacts to solar energy. Each solar cell is very small (about $\frac{1}{2}$ inch by $1\frac{1}{2}$ inches), and therefore, produces only a small amount of electricity. Literally hundreds of millions of these cells will be required in each power satellite. For example, the relatively small solar panels that supply electricity for NASA's Sky Lab needs 147,840 cells to produce 22,000 watts, and these panels weigh over 9000 pounds. The solar energy collecting assembly of the power satellite would weigh over 10 million pounds.

Let's take a closer look at how a solar cell works. All matter is composed of atoms, which are comprised of a central nucleus and one or more electrons that are constantly rotating at high speed around the nucleus (see Figure 68). These electrons tend to remain in their orbits or energy levels. However, when an electron absorbs a certain amount of additional energy, it moves to a higher orbit. It may even completely escape from the atom and become a free electron. Electricity is the flow of electrons through a wire (conductor).

When sunlight strikes the special chemical surface of a solar cell (silicon, cadmium sulfide, or gallium arsenide), the energy of the sun's

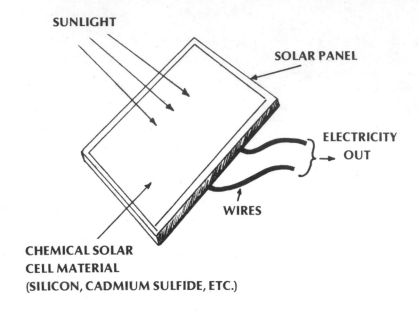

SUNLIGHT

SOLAR PANEL

ELECTRICITY
OUT

WIRES

CHEMICAL SOLAR
CELL MATERIAL
(SILICON, CADMIUM SULFIDE, ETC.)

FIGURE 67. Solar cells work by reacting to the power of sunlight.

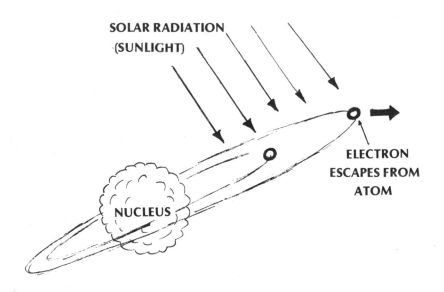

SOLAR RADIATION
(SUNLIGHT)

ELECTRON
ESCAPES FROM
ATOM

NUCLEUS

FIGURE 68. The energy of sunlight (protons) causes electrons to escape from atoms and become free electrons (electric current).

rays is absorbed by the electrons. Some of these electrons escape from their atoms, becoming free electrons and forming an electrical current. Each solar cell contributes its small amount of electricity to form the huge amount of electrical power that is produced by the giant satellite.

Solar cell power supplies in space are not new. They have been in use for many years, providing electricity for unmanned satellites and manned space missions. Most of these solar arrays have been relatively small. The solar arrays on the NASA Sky Lab project, however, were a definite step toward much larger structures. Each of the two solar cell wings for the Sky Lab main power supply measured 28 by 90 feet and contained 73,920 solar cells. The wings weighed a total of 5060 pounds. A second set of solar array panels for the Sky Lab telescope covered an additional area of 1567 square feet and weighed 4221 pounds. The total power output of the four solar arrays amounted to 22,000 watts (about enough electrical power to run a ten-family house). There is no question but that solar power can provide a very meaningful amount of electricity; ultimately, it can supply our total requirement for this type of energy. Why, then, don't we immediately launch a series of these satellites and solve our electrical power problems?

Mainly because there is a big difference between building a highly sophisticated, small satellite costing millions of dollars and building a huge orbiting electrical power station than can compete economically with conventional sources of energy. In the latter case we are faced with efficiency, weight, transportation and reliability problems, plus a host of others.

The first problem lies with the solar cell technology itself. These devices are rather delicate. They are difficult to assemble in large quantities and, most important, they are not very efficient. The solar cells used for satellites in the past convert the sun's energy into electricity with an efficiency of only 10 or 12 percent, at best. This might be contrasted to the 35 to 40 percent efficiency of conventional fossil fuel power plants. Of course, the solar cell is relatively new, having been discovered only in 1954. By using such exotic combinations of materials as aluminum arsenide doped (mixed) with zinc on top of gallium arsenide, scientists at Bell Laboratories have been able to increase the performance of solar cells to an efficiency of 18 percent.

Communications Satellite Laboratories has been hard at work developing a new solar device called a violet cell. According to Joseph Lindmayer, manager of the solid-state physics branch of the laboratories, the violet cell captures the violet and ultraviolet portions of the sun's energy, as well as the visible light energy. Since the violet cells get more of the sun's light to work with, they are able to convert more energy to electricity.

Jet Propulsion Laboratories has claimed the capability of producing

As early as 1954, Bell Laboratories demonstrated the first practical solar cells, using them to power a telephone line. (Courtesy Bell Laboratories)

solar cells with a 19 percent efficiency. And Varian Associates, Palo Alto, California, announced that its new gallium-arsenide cells can operate at a phenomenal 21 percent efficiency. Varian further claims that its new solar cell will reach 23 percent efficiency in terrestial uses. Ronald L. Bell, director of Varian's solid-state laboratory, illustrates the great advantage of the new cell with a practical example. A 10-megawatt solar power plant using conventional silicon solar cells would require a land area of about 24.5 acres; using Varian's new, high-efficiency cells, it would require only 80 square meters. Over the next four or five years it is entirely possible that the solar cell will approach a theoretical efficiency of 35 percent, which would put it in even stronger contention as a major source of electricity. But efficiency is not the only problem.

Cost is another barrier. The low efficiency of solar cells is one reason why this type of power generation is so expensive. How expensive? It is hard to say exactly; there seems to be a good deal of confusion about the actual cost of solar electrical power. Various industrial and scientific journals quote solar cell cost as $2000 per KW all the way up to $100,000 per KW. Even at the low end, electrical energy from solar cells turns out to be very expensive—about ten times more than conventional electrical power. If you wanted to run a common 100-watt bulb using silicon solar cell electrical power, it would cost you between $5000 and $10,000 to keep it lit. So though efficiency is now a manageable problem, the cost has got to come down sharply.

Another cost problem lies with the material itself. Silicon, the most common material used in solar cells, is very abundant and can be bought in metallurgical grade for about $600 per ton. But the extremely pure form of silicon needed for solar cells cost about $60,000 per ton. Scientists are looking to other materials, such as cadmium sulfide, to manufacture solar cells that cost a lot less. However, they are not as efficient as silicon. So scientists and engineers are trying to bring down the cost of manufacturing silicon devices. Reducing silicon waste is one approach. This material usually comes in 1-inch cylinderlike ingots. It is then sawed into very thin slices (about ten-thousandths of an inch thick) which are used to make the solar cells. In the process of sawing the slices a large amount of silicon is lost. An alternative is to produce silicon in a different form—flat sheets, for example—so that no sawing will be necessary and no material will be lost. But this is easier said than done. One cannot just melt a batch of silicon and spread it thin. The silicon used for solar cells must be grown, under special conditions in a high-temperature furnace, where a continuous buildup of a silicon crystal is slowly formed. Tyco Laboratories, Inc., in Waltham, Massachusetts, together with Harvard University, has developed a new process in which ribbons of silicon over 6 feet long can be produced. Normally, this amount of silicon would require the sawing of about 72 slices, and the loss of a corresponding amount of material. The Tyco process, called edge-defined film-fed growth (EFG), could lead to a fantastic reduction in the cost of solar cells, perhaps from about $20,000 per kilowatt to $375 per kilowatt.

Still another technique which has been discovered by engineers at the Three H. Corporation in Wanamassa, N.J., may make it possible to produce silicon solar cells with 95 percent less material. These developments are also helping to solve another problem that causes the price of solar cells to remain high.

Producing arrays of solar cells is now a custom-tailored hand operation, complex and time-consuming. For example, it takes more than 80,000 welded connections to make a 1-KW solar array. Until recently, these welds could not be automated because the arrays have connections on both front and back. But Lockheed has developed a technique that allows all the connections to be located on one side, so the welding operation can be automated. Manufacturing is where the main cost is right now, and the National Science Foundation and industry are working hard to find ways to manufacture solar cells a lot more cheaply. One other important factor is the small size of the solar cell industry. Right now only about 70 KW of solar cells are being produced each year in the United States. A large order for solar cells from the government would draw more big companies into the industry, and their technological capabilities could really help bring down the manufacturing cost. This is only a matter of time, according to P. A. Berman of Cal Tech: "It seems almost inconceivable that such a simple thing as a solar array substrate [the mounting base for the silicon cells] with printed-circuit interconnections and wiring upon which cells are mounted in some simple, economical manner, and over which some inexpensive protective layer is positioned having no moving parts and using no exotic materials, cannot be made for a few dollars a square meter, rather than the thousands of dollars per square meter experienced in the space program. . . ."

The second major hurdle in getting a solar satellite power station into operation is just getting up there. We are talking about a structure that will weigh around 25 million pounds. It would take between 100 and 300 shuttle flights to assemble a single 5000-MW power plant at a cost of about $10 billion. Of course, the cost of installing other satellites would be reduced through experience, mass-production techniques and improved materials and structural design. But it would still cost in the neighborhood of $700 billion to assemble 100 SSPSs with a total capability of 500,000 MW of electrical power. And by the year 2010, 100 SSPSs would be able to furnish only 50 percent of our electrical power needs. But this would be clean, nonvulnerable energy, available to us for virtually forever. We could save on launch costs by simply making the structure a lot lighter, and since the solar arrays represent more than half the weight, they would be a good place to start. One target is to make solar cells thinner. Right now they are about 10 mils thick (10 thousandths of an inch). According to Dr. Peter Glaser, this thickness could be reduced by 90 percent. With the right supporting structure, 1-mil-thick solar cells could operate just as well as present devices. These thinner cells would not only reduce the weight of the satellite and the launch payload, they would also lower the cost of solar cells significantly.

One program called FRUSA (Flexible Rolled-Up Solar Array) was developed for the Air Force by the Hughes Aircraft Company. To hold the solar cells, it uses a very lightweight structure that can be rolled up like a window shade. Instead of getting 1 to 5 watts of electricity per pound of solar ray assembly, FRUSA gets 22 watts per pound. The concept of concentrators, extremely light mirrors that can concentrate additional sun-

RCA technician inspects one of the more than 11,000 solar cells that will provide solar electrical power for the NIMBUS-5 experimental satellite. (Courtesy RCA)

light onto the solar cells, can go a long way in reducing weight. With concentrators, the arrays can get as much as 500 watts per pound of array weight. With additional research and development, the power-to-weight ratio of the arrays will be greatly improved—to perhaps 1000 watts per pound.

Another problem related to the operation of an SSPS is radio interference. The transmission of huge amounts of high-frequency energy may generate a great deal of electromagnetic noise, which will interfere with certain radio frequency bands. Amateur sharing, state police radar and high-power defense radar will suffer. Studies will have to be made to determine the degree of interference and to see if another frequency range can be used, either by communications activities or by the SSPS stations.

When the problems are weighed against the benefits, the scale tips in favor of the inexhaustible, pollution-free energy. How long would it take to put the first of these giants in the sky? Within the scope of present plans and at our present pace of development, many of the basic technology problems will not be solved before the mid-1980s. If the decision to go is made then, long lead items such as adequate launch facilities and the training of hundreds of astronauts would take until the 1990s. The first station might go into operation between 2005 and 2010. But meanwhile, the whole world would have gone nuclear, and we will be hard pressed to justify the huge expenditures required for this "space venture."

On the other hand, with a true concern for the viability of society and some well-balanced logic, we will conclude that a concerted effort to develop a clean, safe alternative energy source to today's depleting fossil fuels is in order. An intensified solar development program could compress the time; the basic technological problems could be solved by the end of the 1970s. Parallel efforts on launch facilities, astronaut training and station structure manufacturing could be completed by the late 1980s, and the first orbiting power station could be in operation by 1995 or sooner.

The concept of an orbiting power station is the brain-child of Peter Glaser. He has summed up the case for this project succinctly in an article published in the August 1973 edition of *Astronautics and Aeronautics* magazine: "Bringing to reality a Satellite Solar Power Station represents a major challenge to engineering and an unparalleled opportunity to apply space technology for the benefit of mankind. But we need no fundamental breakthroughs to achieve this objective. Just as only 15 years ago a space shuttle able to transport large payloads into orbit and repeat the mission 100 times seemed scarcely attainable, so today creating a satellite as large as an SSPS may strike the casual reader as an extraordinarily formidable undertaking. That is really not the point. It can be done. The question space designer and policymaker alike must answer foursquare concerns its relative merit in meeting our needs: Should we create the SSPS? And as we approach this question now, we should bear in mind that the criteria for decision-making based on cost, resource conservation, or environmental enhancement may be quite different in the future, and continue to change as long as technical developments present us with new energy-production methods. We should, therefore, take the necessary steps to protect the SSPS option in the meantime."

Let us now turn to a more down-to-earth approach to using the sun's energy, namely, solar farming. Located right here on the ground, solar farming represents a different technology, one in which solar cells are not the prime device. One might wonder if enough of the sun's energy reaches the surface of the earth to be useful to man. The answer is yes. The earth receives 173 quadrillion (10^{15}) watts of heat energy from the sun each day. The solar energy that falls on Lake Erie alone is greater than the total amount of energy consumed by the entire United States in the same period of time. Every square foot of the United States receives an average of 75 to 104 BTUs per hour in the summer and 21 to 46 BTUs per hour in the winter. This is the most abundant form of energy available to man. The amount of sunlight that strikes the surface of our planet is 7×10^{17} kwh per year. This complex-looking figure is more than 30,000 times all the energy consumed by all the devices (machinery, cars, planes, boats, appliances, etc.) used by man. Scientists have calculated that if we could efficiently collect the sunlight that falls on only 14 percent of the western desert regions of the United States, it could be converted into 1 trillion watts of electricity, more than enough to meet our additional requirements for electricity until 2000.

Most of the sun's power is concentrated in the southern portion of the country, particularly in the Southwest, which has sunny weather about 300 to 330 days per year. The best areas to collect solar energy would be New Mexico, Arizona, parts of Nevada and southern California, southern Florida, southern Louisiana and southern Texas. These huge amounts of solar energy are certainly worth collecting or farming. Dr. and Mrs. B. Meinel of the University of Arizona's Optical Science Center envision solar farms that appear as vast fields of corn in the brightly lit deserts of the United States.

Solar energy has often been thought of primarily as a technology to aid small or backward countries that cannot afford today's high fuel prices. But this "thinking small" is the wrong approach. The Meinels are certainly not thinking small. They envision a series of 1000-MW solar electric farms covering thousands of desert acres and producing pollution-free electricity at prices competitive with conventional electric plants. And as a bonus, the solar plants would furnish billions of gallons of fresh water. The remarkable thing about the Meinels' plan is that it could be put into large-scale operation without waiting for new technological breakthroughs. Before we take a brief look at what takes place on one of these sunshine farms, we should consider some facts about light and dark surfaces.

In hot climates the people tend to wear white because white reflects sunlight very effectively and thereby keeps one cool. Black is less effective in reflecting sunlight; in fact it is a very good absorber of solar energy. A black object left out in the sunlight will get much warmer than a white object. Absorbing the sun's energy is the idea behind solar farming.

One approach to farming the energy from the sun is shown in Figure

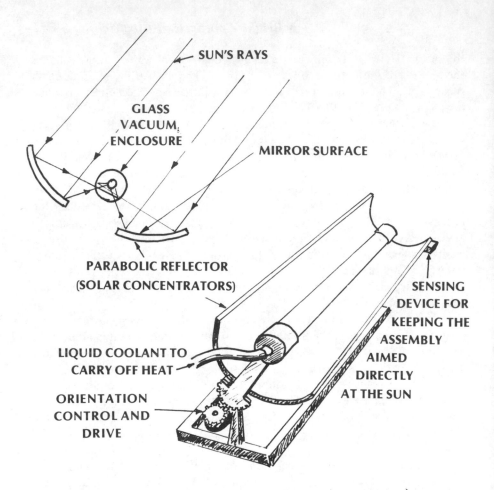

FIGURE 69. Solar collectors using parabolic reflectors (concentrators) can convert sun rays into useful heat energy.

69. Curved metal surfaces called parabolic reflectors (which reflect all the incoming sun rays to one central point) are used to collect the energy. Because the sun is so far away, its rays are considered to be parallel when they reach the earth. The parabolic reflector deflects the rays (changes their angle) so that they all meet at a certain point near the center of the reflector. Located at this point is a steel tube that has been coated so it will absorb the maximum amount of solar energy. An outer glass cylinder encloses the steel tube, providing an area in which a vacuum exists. This prevents heat energy from escaping from the central steel tube. The inner surface of the glass enclosure is coated to reflect sunlight coming from the parabolic reflectors back onto the steel tube. In this way not only does the steel tube receive direct heat from the sun, but it also receives about ten

times the additional solar energy from the parabolic reflectors, which operate as energy concentrators. As a result very high temperatures are generated in the steel pipe (about 1000° F.). This heat is carried off by a cooling liquid (nitrogen gas, air or liquid sodium), which takes the heat energy to a central storage area or directly to a steam generator. Such an experimental unit, mounted on the roof of the Optical Science Building of the University of Arizona, has been going through an evaluation process by Dr. Meinel and his staff.

Here is another approach to an energy farm. Panels about 3 by 10 feet, made of various layers of material, would be built. As shown in Figure 70, each panel would consist of a thin coating of gold (about one-millionth of an inch thick). The coating would be laid on a layer of quartz (a type of glass that can withstand high temperatures). Next, there would be a layer of metal, which would absorb heat. The sun's rays would easily pass through these layers of gold, glass and quartz. The metal layer would absorb much of the heat. The portion of the sun's energy that is made up of infrared rays would reflect off the metal and try to escape from the panel. However, these rays would be reflected by the layer of gold and, therefore, be trapped. The trapping of infrared rays would result in a buildup of a great deal of heat in the panel—the panel would reach a temperature of about 1000° F. The efficient generation of heat is the first and most critical step in this operation.

Another coating that acts as a heat mirror was developed at MIT's Lincoln Laboratory by John C. C. Fan, Thomas B. Reed and John B.

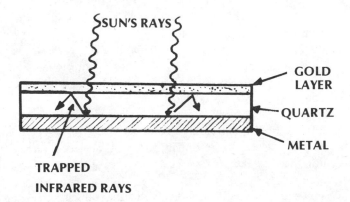

FIGURE 70. Specially constructed panels could collect the energy of the sun by trapping its infrared rays.

Goodenough. It consists of a thin film of gold or silver, only two-millionths of an inch thick on Pyrex glass. Incredible as it sounds, this microscopically thin film has the insulation effect of several inches of asbestos. The coating is transparent to light, but totally opaque to heat or infrared radiation. Such a coating is ideal for solar energy collectors.

A third type of selective coating, developed by Dr. Bernard Seraphin of the University of Arizona, is composed of the element silicon (a semiconductor material), which has natural radiation-selective properties. The silicon semiconductor literally soaks up sunshine and effectively converts it to infrared radiation. Working together with an infrared-absorbing material, it acts like a solar-energy-trapping sandwich. A great advantage of silicon selective coatings is that solar energy collectors can be made about 100 times cheaper than can be made using solar cells. Obviously, such coatings are extremely important to solar collectors, which are built without focusing mirrors or lenses. These are called planar (flat) collectors. Simple planar collectors enjoy a number of advantages over other types of systems. Concentrating systems must always be aimed directly at the sun, so they need motors and control systems, which makes them more expensive and complicated and prone to require service and maintenance. Planar solar collectors do not require this extra equipment and can utilize defuse sunlight as well as direct sunlight. Therefore, unlike concentrating systems, they can operate even on overcast and cloudy days. However, planar systems suffer from a major disadvantage.

Since most selective coatings degrade seriously at high temperatures (above 400° c.), electrical power plants using planar collectors would have to operate at only 350° c. A number of scientists are hard at work developing selective coatings that will operate at higher, more efficient temperatures. A lot of work is also being done on the development of high-efficiency, low-pressure steam turbines. These efforts will certainly improve the efficiency of planar collectors.

Most likely, a combination of planar solar collectors with simple reflecting units (called booster mirrors) to concentrate additional sunlight (about twice as much) onto the collecting surface will be used. One advantage of selective-coated solar collectors is durability.

Since tens of thousands of collectors, spread over hundreds of acres, will be required, it would be extremely costly to have to replace large numbers of panels because they wore out in a short time. So scientists are now weather-testing these surfaces to see how well they hold up against the natural elements.

Many thousands of such solar energy collecting panels (see Figure 71) would be installed. Each panel would have an additional mirror positioned at right angles to the main panel. This booster mirror would reflect

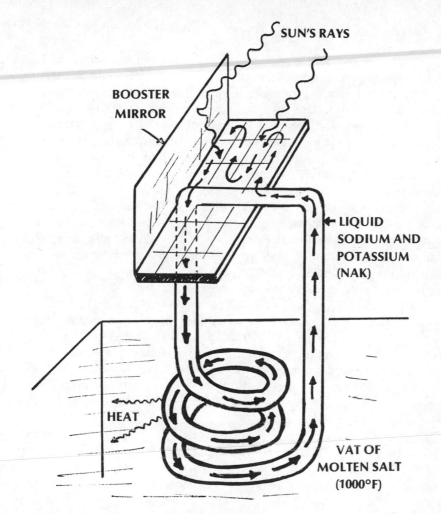

BOOSTER
MIRROR

◄ LIQUID
SODIUM AND
POTASSIUM
(NAK)

HEAT

VAT OF
MOLTEN SALT
(1000°F)

FIGURE 71. Solar energy collector panels would capture heat from the sun and transfer it to a vat of molten salt, which could store heat for two days.

additional sunlight onto the collector panel. The heat in the panel would be collected by a liquid that would flow through a pipe and into the panel and circulate through the panel, picking up heat and then flowing back into a return pipe. This liquid, comprised of sodium and potassium (called NAK), is a good collector of heat. It would be maintained at a minimum temperature of 450° F., so it would remain in a liquid state. When it left the collector panel, it would be 1000° F.

Below the ground, the pipe carrying the NAK would form a coil that would be submersed in a vat of salt. The heat from the NAK would be transferred to the salt. Salt happens to be an excellent substance for storing heat over a long period of time. In fact, these vats of salt would be able to maintain sufficient heat to produce large quantities of electricity even if the sun were obscured by clouds for two days. Of course, there are a number of other alternatives for storing the energy of a solar farm. Under high pressure, water can store a great deal of heat without changing to steam. Therefore, during the day the extra heat not used to generate electricity could be sent to a large tank located underground (for insulation and for added strength to hold in the pressure), which would contain water under pressure. Steam containing the heat could be bubbled through this water, releasing heat to it. During the night or on cloudy days the pressure in the tank could be lowered so that the very hot water would instantly flash into steam and be used to rotate a steam turbine. A major advantage to this system is that one fluid is used to cool and collect the heat from the reflectors, drive the steam turbines and store heat energy.

Energy from solar farms could also be stored in the form of electricity in huge high-capacity batteries. However, such a system would be very expensive if it had to service a 1000-MW installation. Energy could also be stored in mechanical form, using giant flywheels. In most cases meterologists can predict cloud coverage well in advance. Upon notice, the system of flywheels could be spun up with enough energy to carry the plant for a full day. And certainly an evening's use of electrical power could be stored in flywheel units.

Hydrogen and oxygen, produced by electrolysis, could store energy in a chemical form. This method provides a great deal of flexibility; not only could it store extra energy for nighttime and for cloudy days, it could also export solar energy to other parts of the country in the form of H_2 and O_2. Then the Southwest could supply not only all its own energy, but it could also transmit a great deal of energy to other parts of the nation. Incidentally, using the hydrogen and oxygen storage method, a solar farm could accumulate enough fuel to carry it through weeks of totally cloudy days. Pumped storage, compressed-air storage and magnetic storage are some other possible ways to store excess solar energy.

Of course, the excess electricity generated by a solar farm need not be stored at all. It could be transmitted directly into the nation's electrical grid system during peak power use hours. In this way, solar farms would operate as a peaking system. At night and on cloudy days power could be drawn from the national power grid systems. Such a complex operation of give and take using multiple-location power-generating facilities would have to be controlled by large computers.

After the collector, the solar power generating system operates very similarly to conventional electrical generating plants. As shown in Figure 72, also located in the salt vat would be a second coil of pipe containing water. Naturally, since the water coil is immersed in a 1000° F. bed of salt, the water within the pipe would be converted to very high pressure superheated steam. The steam would be directed into a turbine, a jet engine-like device that rotates from the force of the steam. The steam would leave the rear of the turbine and enter a special water-filled tank called a condenser. Here it would be cooled or condensed, then sent back to the salt vat to be reconverted into high-pressure steam. This is called a closed-cycle system. Meanwhile, the spinning steam turbine would drive an electrical generator, which would produce the electricity that is used for homes and factories.

In recent years a lot of research has been done on new concepts in solar farming. One new approach is called the solar furnace. It is presently

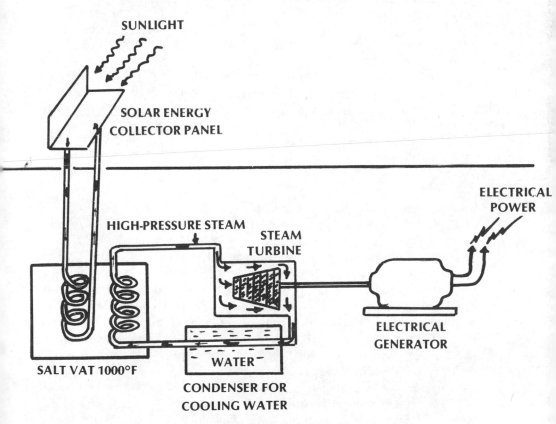

FIGURE 72. Electricity-generating system powered by heat from solar collectors.

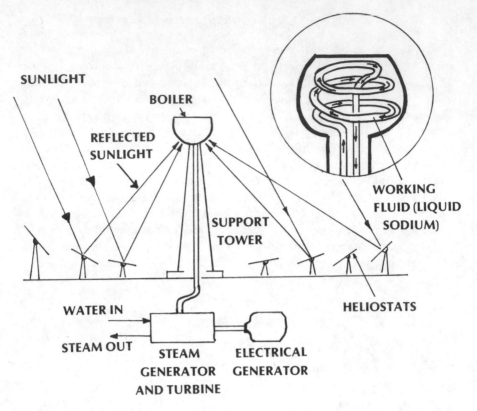

SUNLIGHT

BOILER

REFLECTED
SUNLIGHT

WORKING
FLUID (LIQUID
SODIUM)

SUPPORT

TOWER

WATER IN

HELIOSTATS

STEAM OUT

STEAM
GENERATOR
AND TURBINE

ELECTRICAL
GENERATOR

FIGURE 73. A central energy-heating (using a boiler) concept to produce electricity from the sun.

being studied by a team of scientists from the University of Houston, McDonnell-Douglas Astronautics Company, Martin Marietta, Georgia Tech, Sheldahl, Inc., and Foster Wheeler. The solar furnace (see Figure 73) is based on Archimedes' idea of over 2000 years ago—that is, concentrating the reflected sunlight from many mirrors onto a single point. In the solar furnace the single point is a huge steel boiler that has been covered with selective coatings so that it will absorb the maximum amount of solar radiation. The height of the tower holding the boiler would be about 770 feet. Spread out all around the tower would be a large number of reflecting mirrors (over 1800 of them), called heliostats, measuring 20 by 20 feet. These heliostats will be rotated on two axes so they can track the movement of the sun. They will also be positioned so that each of them will reflect sunlight onto the boiler (which measures 47 feet high and 37 feet in diameter) at the top of the tower. Obviously, with so much sunlight concentrated on it, the boiler will become very hot—about 650° c. Within

This huge central boiler will collect solar energy concentrated on it by hundreds of mirrors and use the energy to change water into high-temperature steam. (Courtesy Martin Marietta)

the boiler is a series of coils that will also get very hot. A liquid such as hot sodium or a combination of sodium and potassium (NAK) can be used to carry the heat from the boiler down the tower and into a steam generator, which provides the steam to drive a turbine. Also, water could be pumped directly to the boiler, where it will be changed to high-pressure, super-heated steam, which would be used to drive a steam turbine directly.

Scientists are also considering the use of solar cells as collector elements on solar farms. Such an approach would not require a cooling system or steam-generating equipment or even an electrical generator—fuel cells would change sunlight directly into electricity. Right now, this system is far too costly to build, but once manufacturing techniques improve, solar cell efficiency is raised and mass-production methods bring down the price of solar cells, such a power system will be very practical.

It is possible that a hybrid system of solar collectors will emerge in which the best qualities of selective coatings and of solar cells are combined. Figure 74 shows such a design. It is a solar sandwich in which the top section consists of selective coatings that trap infrared (heat) energy

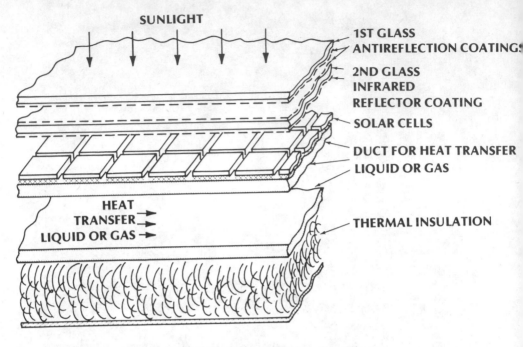

SUNLIGHT

1ST GLASS
ANTIREFLECTION COATINGS

2ND GLASS
INFRARED
REFLECTOR COATING

SOLAR CELLS

DUCT FOR HEAT TRANSFER
LIQUID OR GAS

HEAT
TRANSFER
LIQUID OR GAS

THERMAL INSULATION

FIGURE 74. A combination thermal-photovoltaic solar collector.

while allowing the visible light of the sun to pass through and strike the solar cells. The heat is carried off by a coolant and used to generate electricity. Meanwhile, the solar cells generate electricity directly.

Whatever method eventually proves best, we will someday see vast plains covered with acres upon acres of solar collectors. How much land will a solar farm require? Estimates vary, but they range from about 3 to 20 square miles per 1000 MW of power. The Meinels envision a series of solar farms across the southwest desert in a corridor along the lower Colorado River. These power plants would supply 1 million MW of electricity, enough to meet total U.S. demand in the year 2000. The land covered by these farms would be equal to a square measuring 74 miles on each side. This may sound like a big area, but it is less than 1 percent of the land we presently use for farming. And, it will be arid desert land. Furthermore it would require no strip mining, no dangerous deep mining, no accumulation of deadly radioactive waste and no air pollution. Furthermore, it should be noted that this amount of land is far less than what is presently being devastated by strip mining. It is a small price to pay for all the benefits solar farming has to offer.

Cost estimates also vary greatly. Dr. Meinel is now trying to raise enough money to build a small working pilot plant that would operate in the desert. It would cost about $4 million. It has been estimated that large solar plants using thermal methods (as opposed to solar cells) would cost about two or three times more to build than would fossil fuel plants, or $1000 to $2000 per KW. But the fuel for solar plants is free. And as fossil fuels become scarcer, the cost advantage will swing to solar power. One element that may keep the cost of a thermal solar farm high is materials; copper, aluminum and steel, required in great quantities for solar farm construction, will most likely continue to rise in price. The cost of solar cells will have to be reduced about 1000 percent before they will be economically eligible. Eventually, thermal solar farms will produce power for about $600 to $1000 per KW, which will be very competitive with other power-generating methods (nuclear plants are quickly approaching the $700-per-KW level).

Here is an example of how solar energy can actually save the United States a great deal of money while solving our energy and pollution problems. The average amount of sunlight that falls on a square foot of land in the United States is approximately twice the amount needed to heat and cool the average house (i.e., 17 thermal watts per square foot). Converting this energy into electricity at a 10 percent conversion efficiency would result in an average daily electrical output of approximately 1140 MW-hours per square mile. In 1969 the Potomac Electric Power Company sold a daily average of 30,000 MW-hours to 425,000 customers in an area encompassing 643 square miles. If only 4 percent of the area serviced by PEPCO were devoted to solar generation, the utility could have provided all the electricity that was required. This is not to suggest that all communities can afford to use 4 percent or even 1 percent of their available land for power generation. In certain parts of the country where real estate is at a premium, such a concept would be economically unfeasible. But where the land is not in constant use, it can economically be employed for electrical power generation.

Some people have claimed that solar farms would not really be pollution-free because they would change the heat balance of the area where they were situated. This is not quite true. Ordinary desert land absorbs approximately 65 percent of the solar radiation and reflects 35 percent back into space (and into the atmosphere). If a solar farm were installed, the area would absorb 95 percent instead of 65 percent, a difference of 30 percent. But 30 percent of that energy would be transmitted to electrical grid systems in the form of electricity. So the energy concentration in the area would be the same as it was before the solar farm was installed. But the farms would be another source of excess heat. You may recall that when steam leaves a turbine, it is cooled back to its water state. In the process a certain amount of heat must be carried off from the system. What do we do with it? It could be a form of heat pollution. But remember, we are in the desert where there is an extreme lack of water. Therefore we can turn our heat pollution into a benefit. Salt water could be pumped in from the ocean (in the southwestern United States, from the bay of California) and converted by evaporation into fresh water using the excess heat from the condenser. Large quantities of fresh water could thus be made available to the desert region. In fact, a solar farm designed to produce 1 million MW of electricity could also produce as much as 50 billion gallons of fresh water per day, enough for a population of 120 million people. Such a vast amount of fresh water could eventually turn a barren desert into rich, fertile farmland and a pollution-free vacation area. It would also be possible to reduce excess heat significantly by using a low-temperature bottoming cycle to produce electricity from the waste heat.

What is the present status of solar power? What are the problems to be solved? No new technologies have to be developed. No new major breakthroughs are required. There are engineering problems, materials problems, manufacturing and cost problems. But none of these is a major barrier in the path of progress. That is why Dr. Meinel stated to a congressional committee in the spring of 1972 that the timetable for solar energy could be the same as that for the projected atomic breeder-reactor program—which is so strongly backed by government and private industry. There is little doubt that a 500-MW demonstration solar plant could be in operation by the mid-1980s at a cost less than that for the breeder program.

Another method of using solar energy has been proposed that does not require huge structures in space or large land areas. The entire collection and processing system would, instead, be located out at sea. This, the Helios-Poseidon method, has been proposed by a technical consultant named William Escher of St. John's, Michigan. It involves huge square arrays of solar cells (about a half-mile square) mounted on large floating platforms situated on the ocean at the equator.

Another, similar approach has been proposed. It starts off with the same basic idea. Solar energy collectors would be used to gather the sun's energy. However, these would be disclike reflectors that concentrate the sun's rays on a point at the center of the reflector (see Figure 75). The center or focal point would become so hot that it could raise the temperature of water coming into contact with it to 600° c. At this temperature the water would be turned into superhot, high-pressure steam. Figure 76 shows how the complete system would work. The steam created by the solar energy collector would be used to run a steam turbine. The steam turbine, in turn, would drive an electrical generator. Once the steam passed the turbine, it would be pumped through pipes to a cooling condenser located

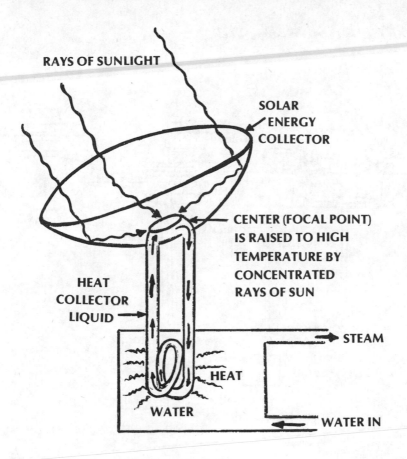

RAYS OF SUNLIGHT

SOLAR ENERGY COLLECTOR

CENTER (FOCAL POINT) IS RAISED TO HIGH TEMPERATURE BY CONCENTRATED RAYS OF SUN

HEAT COLLECTOR LIQUID

STEAM

HEAT

WATER

WATER IN

FIGURE 75. Focal-point solar energy collector system.

deep below the surface of the water. At this depth the water might be as cold as 4° to 7° c. Therefore it would rapidly cool the steam, returning it to water.

Because this solar power station would be located in the tropical seas, it would be highly impractical to transmit the power to a populated area. Therefore the electricity produced by the generator would not be used for general consumption but to operate a special device called an electrolyzer, which would be located about 3000 feet below the surface of the ocean. In the electrolyzer water would be separated into its basic elements, oxygen and hydrogen. The oxygen and hydrogen would then be pumped into flexible or thin-walled tanks. The advantage of having the storage tanks located at this depth is that here water pressure is sufficient to keep the hydrogen and oxygen gases under high pressure without constructing very strong and

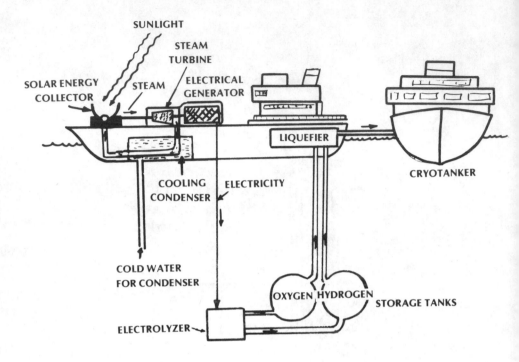

FIGURE 76. Floating solar power station.

very expensive storage tanks. Furthermore, this type of storage would be very safe; the tanks could not burst because of gas pressure because the surrounding water pressure would be equally high.

These gases would become a source of fuel for fuel cells or gas turbines in conventional electrical power plants around the country. Whenever fuel is required, the gas would be directed to the surface, where it would be liquefied and pumped into a special cryogenic tanker (cryotanker), which would then deliver the liquid gas to an electrical power station.

One disadvantage of this system is its susceptibility to damage during fierce storms. Still, this type of solar farming would not take up the large amount of property required by a land-based system, and it would be more efficient since the extremely cold water needed for cooling purposes is readily available. There would also be a cost savings in storing the oxygen and hydrogen produced, since the tremendous weight of the ocean would provide the pressure to support an inexpensive thin-walled container. Nor would there be any thermal pollution, since the ocean is an infinite heat sink (i.e., capable of absorbing a tremendous amount of heat).

One novel method of using solar energy offers a significant advantage over all the previous methods: it can operate for a considerable period of time even if the sun is hidden by dense cloud cover. This method requires that the entire generating plant be located about 200 feet below the surface of the water in the tropics. The idea of extracting power from the sea is based on the fact that the oceans contain a vast amount of energy—in fact, they are already acting as a tremendous storage tank for energy. Although they receive some heat from within the earth, the oceans get most of their heat energy from the sun—just 1 cubic mile of warm seawater has absorbed literally trillions of BTUs of heat energy. The total amount of heat stored permanently in the oceans is equal to the total amount of solar radiation that falls on the earth over a period of 146 days. This comes to a staggering 2.2×10^{17} MWH or 2200 quadrillion MWH of power. The Gulf Stream alone contains over 75 times the total power production of the United States. A NASA study has shown that 100 percent of United States energy needs in 1985 could be met by Gulf Stream solar energy generators. This sounds impressive, but how does one go about getting electricity out of warm ocean water?

Georges Claude, a French physicist who invented the neon lamp, came up with a novel idea. He performed a number of experiments off the coast of Cuba in the 1920s, but various technologies had not advanced enough for his scheme to work. He was proposing that we take advantage of the different temperatures that exist at various levels of the ocean. In 1969 John Isaacs calculated that with Claude's method, it would be possible to extract 200 times more energy than man could ever use. In 1930 Claude had proved the basic concept by sinking a shaft 1800 feet into the Matanza Bay in Cuba, and generating enough steam to run a 22-KW generator. But the net power gain of the system was very small. He was faced by a number of problems, most of which have since been solved. How does ocean thermal differential power work? Between the Tropics of Cancer and Capricorn, where the intensity of incoming solar energy reaches its peak, 90 percent of the earth's surface is water. The surface layer of this water never drops below 82° F. Meanwhile, to the far north and south (at the earth's poles) the sun's energy melts the ice and cold water slips down to the bottom of the oceans. This cold water slowly moves toward the equator, forming the deep cold-water ways of the ocean. As a result, over an area of several hundred million square miles there exists a level of very cold water and level of very warm water within a few thousand feet of each other (the separation varies at different locations.)

The surface temperature is about 85° F., the lower level temperature, about 45° F.—a difference of about 40° F. This difference is the key to generating electrical power. There are two similar methods of extracting electricity from this source. The first one, called the Claude system, is shown in Figure 77. Warm water from the ocean's surface is pumped into the top of a long cylinder called the expansion chamber, which is the heart of the system. The water enters a small area at the top and pressure forces it through a narrow passageway called an orifice (the action is somewhat similar to a water pistol forcing a stream of water through a tiny hole in the front of the gun). As it exits from this orifice, the water enters a large chamber. This rapid change from high pressure to a low pressure causes

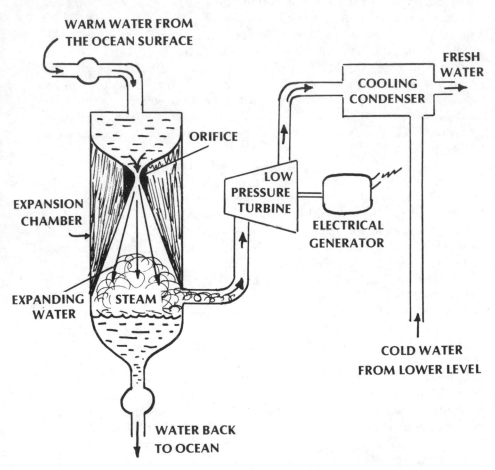

FIGURE 77. The Claude System for extracting electricity from ocean thermal differentials.

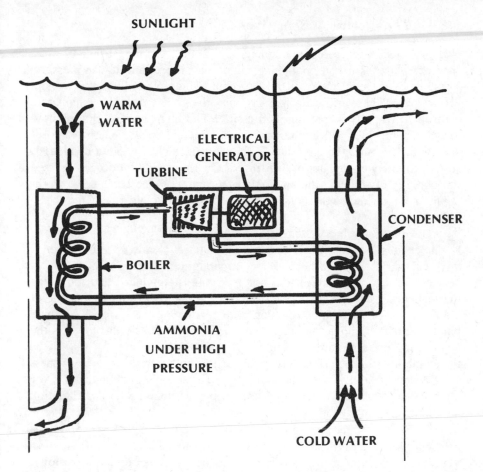

FIGURE 78. Solar heat in seawater would be used to expand ammonia (or another gas).

the water to expand, separating into vapor (steam) and water. The water continues to fall to the bottom of the chamber; the steam travels through a pipe into a low-pressure steam turbine. The steam rotates the turbine, driving an electrical generator to produce electricity. The steam then passes to a cooling condenser, where cold water from the ocean 2000 feet below is brought up to cool the steam and change it back to water. However, the water is now fresh, having lost its salt when it was turned into steam; so the system not only generates electricity, it also produces large amounts of fresh water.

In the early 1960s a study was made by James H. Anderson and his son J. Hilbert Anderson of another type of ocean thermal power generator. Figure 78 shows this method. Warm water from the surface enters the top of a large steel tube 35 feet in diameter and 2000 feet long. It then enters a chamber that has a coil containing ammonia or propane under pressure. These fluids are extremely sensitive to heat, which causes them to expand very rapidly. The warm water passing over the coils expands the ammonia or propane, rapidly increasing its pressure. The high-pressure fluid is used to drive a turbine, which, in turn, rotates an electrical generator. After driving the turbine, the ammonia is directed through a cooling condenser. Here, extremely cold water drawn from a few thousand feet down is passed through, rapidly cooling the ammonia or propane. Once cooled, the working fluid is returned to the boiler, where it is reheated and redirected to the turbine.

Obviously, such a system uses no fuels at all. Furthermore, since it has no exhaust, no smoke, no radioactive material or any form of effluents, this generating system would create no pollution. It is a system almost perfectly in turn with nature. But not everyone is convinced that the ocean thermal energy can be a practical source of power. Critics have pointed out a number of technical and economic problems facing these solar sea power plants (SSPP), although nobody has questioned the basic theory. It is the engineering tasks required to build commercial plants that are said to be the stumbling blocks; critics also point to the high degree of corrosiveness of seawater, insufficient temperature difference, fouling of machinery by microbes in seawater, reliable sea anchoring, diluteness of solar radiation and the effects on the environment.

For years the effects of seawater on structures have been a serious problem. Because of the electric conductivity, metals that are submerged in seawater have a tendency to corrode and decay (electrolytic corrosion). Metals such as iron corrode rapidly in seawater, while metals such as copper, silver and gold, while highly resistant to corrosion, cost far too much. However, one metal has the unique properties that make it a perfect candidate for this job. Aluminum forms a thin layer (an oxide coating), which seals and protects it from the effects of seawater. Pure aluminum would not be strong enough for an SSPP structure. But bonding a thin layer of pure aluminum to high-strength aluminum alloy would solve the problem of seawater effects on the SSPP.

The problem of insufficient temperature difference is mainly a mental one. Most power engineers are accustomed to working with high-temperature systems where the temperature difference between the heat source and the turbine exhaust is about 500° F. The SSPP method, with a temperature difference of at best 86° F., is rather startling to them. But two factors make the SSPP method workable and practical: the tremendous amount of water being passed through the system and the use of special fluids that have a high-vapor pressure at ambient temperature. Adding a small amount of heat will raise the pressure enough to drive a specially designed

Certified Sun—a NASA Lewis Research Solar Simulator testing a new solar collector design. (Courtesy NASA)

turbine. Furthermore, in conventional high-temperature plants the pressure of the steam is very high (about 3200 PSI), so the boiler walls must be very thick and strong. In atomic power plants heat must pass through thick walls in order to be transferred to the working fluid (water), making plant costs very expensive and reducing efficiency. In the SSPP the working pressure is only 150 PSI, so very thin boiler walls can be used, resulting in a less expensive and more efficient system that uses more of the heat.

The next criticism to be answered is that microbes will foul up the machinery. The surfaces of structures submerged in seawater quickly assume a covering of microbes, which would hinder the transfer of heat from the boiler to the ammonia or propane working fluid. But scientists at the Woods Hole Oceanographic Institute have solved this problem: by adding extremely small amounts of chlorine (less than one part in four million), they can prevent microbial growth. The concentration of chlorine used is far below that needed to kill marine life; when the chlorine is added to the water entering the SSPP and a small electric current is passed through the water, hypochlorous acid is immediately formed, which prevents the microbes from growing.

An SSPP would be a huge structure, and the problem of anchoring it would be difficult, though not insurmountable. Since the structure would be buoyant, the tiedown would only have to keep it from drifting because of undersea currents. With the proper site, this problem could be minimized.

The criticism that the solar radiation is too diluted and that plant efficiency is low can be answered somewhat similarly to the criticism that there is too small a temperature differential. No fuel has to be paid for, so the final cost of the plant and its level of power production are important factors. A number of scientists have estimated that power produced by an SSPP would be only one-third as expensive as that produced by conventional fossil-fueled plants and one-fourth as expensive as atomic power. In fact, Lavi and Zener have estimated that SSPP electricity would cost less than $\frac{1}{3}$ that from fossil fuel and about $\frac{1}{4}$ that of an atomic plant.

Ocean thermal power systems create no pollution problems; the effects on the environment balance would be negligible. There is concern about lowering the temperature of the tropical oceans by mixing large amounts of cold water with warm tropical waters, but this could hardly be a problem. If the entire amount of electricity consumed by all the people on this planet in the year 2000 were supplied by SSPPs, the ocean temperature would be lowered by only $1°$ C. Therefore power stations supplying 10 percent of U.S. needs would cause no measurable temperature change in the oceans.

Neither should markets for SSPP electricity be a problem. With power plants located reasonably close to the coast, electricity could be fed to the mainland via buried cables; with power plants located at farther distances, energy in the form of liquid hydrogen in tanks could be shipped anywhere. Furthermore, some industries such as aluminum could go to the source of cheap electricity. Most of the bauxite used in America comes from Jamaica. With cheap electricity from SSPPs, aluminum companies could locate their manufacturing facilities right in the Caribbean, thus eliminating the need for shipping tons of ore.

Here we have a great method of producing huge amounts of electrical power, cheaply and with absolutely no pollution. Is something being done to take advantage of it? Until the energy crisis, the SSPP concept was just that, an idea, a scientific and engineering curiosity. But some eminent scientists have strenuously proposed that we seriously consider ocean thermal power as a partial solution to the energy crisis. The government is also beginning to participate. The conceptual plans for an ocean heat power plant are presently being subsidized by the National Science Foundation at a cost of over $1 million. They call for an SSPP, possibly as large as 400 MW, to be located in the Gulf Stream off the Florida coast, about 15 miles east of the University of Miami. Such a plant could support the electrical needs of a city of 400,000 people. Research in this area is also being performed by the Naval Undersea Center at the Scripps Institute of Oceanography in San Diego, California, and at the Naval Civil Engineering Laboratory in Port Huenene, California. Because of its universal availability and because of the many ways in which it manifests itself, solar energy research and development is beginning to be recognized as the real technology of the future.

One area of solar energy, which had been associated only with growing things, is now being studied as another avenue to the huge reserve of the sun's power—photosynthesis. Photosynthesis is that strange process by which plants convert the light from the sun and water from the ground into chemical energy to build leaves, roots and other parts of vegetation. Scientists refer to the process of producing fuel from plant life and other living matter as bioconversion. Most growing things can provide an excellent source of energy through the use of photosynthesis.

One method is anerobic digestion by algae (bacterial fermentation) of sewage, garbage and vegetable matter, which produces methane, a good fuel gas that can be compressed and burned in engines or used to heat homes. Hydrogen can also be produced with this method. Using the sun's energy and animal waste to provide the bacteria, algae can be grown on ponds and harvested; the collected algae can then be fermented and used to produce a fuel gas that consists of up to 70 percent methane. About 5 cubic feet could be extracted from every pound of algae. Thus a great deal of fuel gas, to supplement the nation's dwindling supply of natural gas, could be produced. Unlike other synthetic gas processes, algae growing would inject no poisonous gases into the air. Bioconversion is becoming very popular with farmers and livestock breeders. Animal manure is fed into a large tank called a digester (which is sunk into the ground); through a type of fermenting process, the manure is biologically decomposed, which results in the production of methane. The solid material that remains in the digester is used as fertilizer.

The sun stores a tremendous amount of energy in vegetation. Why not use it as fuel? According to G. C. Fvego, president of Inter-Technology Corporation, by burning plants, it is possible to attain higher temperatures than can be attained with solar collectors. Living vegetation can be burned to generate large volumes of high-temperature steam to operate electrical generators. As Fvego explains, "We could cultivate a crop that gives us a premium fuel from a perpetually renewable source. . . . Ideally, it would mature in three to five years." Such a crop could yield 200 million BTUs of heat energy per acre. It has even been suggested that all the world's energy requirements in the year 2000 could be met by burning high-energy plants, and all this vegetation could be cultivated on only 4 percent of the world's land area. In the United States, energy plantations covering about 400 square miles (open for recreational use) could be established. Areas where trees and plants grow rapidly would be chosen. Tree trunks, limbs, roots and plants could be burned to fuel a 1000-MW electrical power plant. A less ambitious variation on this idea is to use the waste from present forestry operations (wood chips, branches, stumps, roots, etc.) as fuel. It has been estimated forestry waste could have supplied between 10 and 20 percent of U.S. energy needs in 1975. The main problems with this concept are finding cheap land that can sustain the rapid growth of high-energy vegetation, and reducing the cost of collecting and transporting the material to power plants. Otherwise, no new technologies are needed. Neither scientists nor the federal government has suggested that we attempt to supply the nation's total electrical needs by burning vegetation. But this is one of a number of alternatives that, if combined, could meet the energy needs of our society without endangering its political or social stability or the health and welfare of its citizens.

The first phases of solar energy will probably be addressed to heating and cooling private homes. In its present state of development this is the most practical and efficient use of the sun's power. An average one-family house in the northeastern section of the United States requires approximately 35,000 KWH of heat energy per year for heating. It should be feasible to supply this amount of heat from solar collectors. Hot-water heaters use about 4000 KWH per year, so there should be no problem collecting enough solar energy for that purpose. Although not too many private homes are using solar energy and therefore an accurate estimation is difficult, experts say that it would cost about $3000 to install a solar system to heat and cool a typical three-bedroom house—a pretty good price since solar systems require very little maintenance. Many scientists are convinced that solar heating and cooling of single-family homes will be common by the end of the 1970s, and that solar power will be a lot cheaper than fossil-fueled systems. But the initial high cost of the systems might slow up progress (even though the fuel is free). Developing the technology is not enough; the building industry has traditionally been slow to implement new concepts. Unless the federal government provides tax incentives, consumers will be discouraged from buying new homes that carry the price tag of a solar system.

Some, however are willing to pay the price. Many people in California and Florida are using solar systems to heat their swimming pools; some have graduated to heating their entire houses. According to one survey, there are already nearly 200 solar-heat houses built or under construction in the United States. Interest continues to increase. Harry Ghomason, a Maryland patent attorney who is also an inventor, has been designing solar energy homes for the past 15 years. He has sold about 2000 blueprints of homes similar to his own. He spent about $2500 for his heating system, which provides up to 90 percent of his winter heat requirements. He also has a back-up oil heating system in the basement.

In Columbus, Ohio, an experimental four-bedroom home with solar panels on its roof and a computer in its garage was on display at the Ohio State Fair. The home was built, with the aid of Ohio State University, by Homewood Corporation, which intends to market similar homes in the $30,000 to $50,000 price range in the near future. An experimental house called Solar One built by the University of Delaware has been designed to provide a great deal of technical information, answer a number of questions about solar systems and demonstrate the practicality of some present designs. Situated on the university's campus, Solar One will draw 80 percent of its energy needs from the sun (imagine eliminating 80 percent of your fuel and utility bills). On the roof facing south are two large solar collectors made of the latest cadmium sulfide solar cells. The basement contains a special heat storage system for when cloudy weather sets in. Would such a house be very expensive? According to Dr. Karl Boer (a well-known solar energy scientist), a home based on the Solar One design would be about 15 percent more expensive than a conventional home. But he hopes to reduce the cost of the solar-generated electricity to 3 cents per KWH, which is close to the 2.7 cents per KWH national average (based on 1974 prices). This experimental house was conceived to be evolutionary; it will attempt to bridge the gap between the totally energy-independent home and the power utility company. In fact, Solar One is connected to the local power grid in such a way that it can supply electrical power to the local utility in an emergency.

The idea of energy independence is really taking off. A builder of luxury homes east of San Diego (Caster Development Corporation) has installed a solar collector on the roof of one of its new homes. Any unused heat is stored in a specially designed storage unit for use in evening hours. Caster expects the $1500 solar system to reduce natural gas consumption by 50 percent, so the system should pay for itself in less than five years. But for enthusiastic response, the city of Colorado Springs would win the prize. The town's bankers, builders, educators and council are looking into all the possible ways of using the sun's energy. This started when the citizens of Colorado Springs experienced the trauma of a moratorium on new natural gas installations because the utilities were not able to guarantee an adequate supply of fuel. In December 1973 Mayor Andrew Marshall organized a citizens' group to study the problem. Thus was born the Phoenix Project, which involved the construction of an experimental solar home.

The $95,000 home was financed by the local banks and the Homebuilders Association, and the National Science Foundation contributed $46,000 for electronic monitoring and analysis equipment. The project was a complete success; the house gets 85 percent of its heat from the sun and natural gas provides the additional energy. Today a second solar home stands across the street, fully paid for by its owner and a $30 million hospital is being built which will incorporate a $1 million dollar solar heating system. According to Douglas Jardine, one of the leaders of the Phoenix Project, the next step will be neighborhood solar energy systems; he plans to heat 54 low-income homes from a central solar collection system. Mayor Marshall says that by 1980 as many as 20 percent of all the homes in Colorado Springs could be heated by the sun.

The idea of a solar-powered community is beginning to grow. Sandia Laboratories, for example, is engaged in a program that may soon lead to a total solar energy system for a community of as many as 1000 homes in Albuquerque, New Mexico. One of Sandia's studies shows that solar radiation could supply 60 percent of Albuquerque's energy needs.

Home-based solar systems are just the start; commercial areas, shopping centers, department stores and office buildings are all good candidates for such systems. Companies such as Standard Oil (Ohio), Aluminum Company of America, PPG Industries and the Oliver Tyrone Corporation are engaged in bringing solar power to commercial buildings.

In Lincoln, Massachusetts, the Audubon Society has built a two-story solar energy building, the first commercial building that is both heated and cooled by the sun's rays. It contains 8000 square feet of office space, library and research areas. Its 45-degree roof is formed by a 3500-square-foot solar collector.

In McLean, Virginia, the Mitre Corporation is investing its own money to fight the energy crisis by installing a rooftop solar cell power system that will generate 1000 watts of electrical power, the world's largest ground-based solar cell installation to date. It is also spending $130,000 for an electrolysis system to convert solar electricity into hydrogen, which will provide an energy storage capability.

In Valley Forge, Pennsylvania, the country's first private industrial-scale solar-powered heating system was dedicated at General Electric's Valley Forge Space Center. This experimental system will supply about 75 percent of the heat and hot water used in the huge (20,000 square feet) cafeteria-kitchen area. It is estimated that the system will save this GE facility nearly 12,000 gallons of fuel oil during an average heating season.

In Hampton, Virginia, the National Aeronautics and Space Administration is building a large solar-powered system for a new one-story office building at its Langley Research Center. A 15,000-square-foot solar collector will heat water to over 200° F. to provide energy for heating and cooling. If it works well, the collector will be expanded to about 40,000 square feet, which would make it large enough to supply heating and cooling for all the buildings at the research center.

These activities are just a start. In a special energy study performed by the TRW Company (under the auspices of the National Science Foundation) solar energy is envisioned as a $1 billion industry by the year 2000. By that time 4 percent of all buildings in the United States will be heated and cooled by the sun. Another study, conducted by the research division of Arthur D. Little, projects that the solar power market could expand to $1.3 billion by as soon as 1985. When potential manufacturers reflect that sophisticated and highly pragmatic businesses are going to solar power, they will not hesitate to enter the market. RCA's 70-story tower in New York City, for example, will have a solar power generating system constructed on its 12th-story setback. A 5000-square-foot flat-plate solar collector will provide about 15 percent of the center's air-conditioning requirements and about 20 percent of its heating needs. Even more impressive are the plans for the new Citicorp Center in Manhattan; the entire roof of this 56-story office building will be covered with a solar energy collection system to provide clean energy for the building's heating and cooling systems. As markets appear, so will manufacturers. Already there are over 100 small companies making solar energy equipment. As the markets grow, the big corporations will reach for the big gold ring. By 1980 sales of solar cells alone could reach $400 million; with the total solar power market eventually reaching more than $25 billion, it's a pretty big gold ring. But before the market actually develops, there are a number of areas where research could help a lot.

The effort is really worth it. If solar power could supply only 1 percent of U.S. energy needs by 1983, according to Dr. Peter Glaser, we would save about 100 million barrels of oil each year. He goes on to state that by 1993 the savings from solar cooling and heating systems could equal 2 million barrels of oil each day. To encourage such activity a number of companies including Alcoa, Ashland Oil, Corning Glass, du Pont, Florida Power & Light, FMC Corporation, Ford Foundation, Honeywell and United Technologies are supporting a project initiated by Arthur D. Little—establishing a climate control industry. It will identify bona fide business opportunities in solar climate control, evaluate actual hardware and assist new business ventures in the field.

Scientists in industry and in universities are busy trying to make solar power a reality—practical, economical and competitive. Unfortunately, the United States has lost time because there has not been enough research money and national support. For years practically no one supported solar power projects. In 1970 U.S. electric utilities spent less than ¼ of 1 percent of their gross revenues on research of any kind, and of the 2000 research projects that were being conducted, none dealt with solar power. NASA has been the main support of solar research and development; it spent about $40 million from 1966 to 1971 on solar cells, mainly for space projects. To the federal government solar power was an unwanted guest. For about 20 years it doled out a meager $100,000 per year on solar power projects. In 1972 federal support amounted to only $4 million (as opposed to $250 million for nuclear power), in 1973 solar power funding crawled up to $13 million and to $32 million in 1974. But what is needed is high-level federal funding, on the scale of the nuclear energy projects. Dr. Walter E. Morrow, Jr., associate director of MIT's Lincoln Laboratory, proposes a program that would dwarf the moon-landing program budget. He suggests that the United States (including private industry) spend about $300 billion between now and the year 2000. With such an effort solar power could supply 13 percent of all energy consumed by the end of the century. And by the year 2020 solar power would account for 26 percent of all energy consumed in the United States. This is a staggering amount, equivalent of about 4 billion barrels of oil annually, and there is no way on earth that the United States could get that much oil for its own use. Unfortunately, the federal government is still ignoring the advice of many eminent scientists. Its energy budget for 1976 is $1.604 billion. Of this, solar and geothermal power (both capable of solving the energy crisis) will get a mere 7 percent. Fossil fuel research is to get a full 19 percent and, to the horror of many concerned citizens, nuclear power research will receive a whopping 28 percent (about $440 million). The reaction of Representative Charles A. Vanik (D-Ohio) the ERDA energy budget breakdown was very descriptive: "mostly just a retread of the old Atomic Energy Commission programs. It provides a miserable pittance for the research and development of non-nuclear energy sources." In its long-range energy program, ERDA will spend $10 billion between 1975 and 1979. Of that amount over 20 percent will go for fossil fuel research and 40 percent for atomic power research. Only 10 percent is slotted for solar power. It appears that the U.S. is fully committed to its mindless plunge into the devastating jaws of the nuclear monster and to using up any remaining supplies of fossil material left, regardless of the disastrous effects it will have on the non-energy industries that rely on this raw material.

There are a couple of developments that do offer some hope for the eventual triumph of solar energy over less acceptable power sources. The National Science Foundation has plans to build a 100-MW central solar thermal power plant (similar to the one illustrated in Figure 73). Says Dr. Harold Spuhler (NSF's program manager for the Advanced Energy Research and Technology Division), "Provided that operational costs are acceptable to utilities, we think that 40,000 MW of electrical installed capacity by the year 2000 is a realistic goal to shoot for. . . . With an accelerated program and some form of incentive, we could double that output to 80,000 MW of electrical installed capacity."

A second ray of hope is the Solar Heating and Cooling Demonstration Act of 1974. A total of 4000 homes (half of them private) are to be equipped with solar heating systems, some with cooling devices. A number of public buildings will also be so equipped.

As far back as 1972 NSF and NASA organized the Solar Energy Panel, comprised of nearly 40 top scientists and engineers possessing expertise in solid-state physics, chemistry, microbiology, power engineering, architecture, photovoltaics and the thermal sciences, as well as several economists, environmentalists and sociologists. They were charged with the task of assessing the potential of solar energy as a national energy resource. Their conclusion? "The Panel is confident that solar energy can be developed to meet sizeable portions of the nation's future energy needs." The following are the conclusions and recommendations by the panel as published in a special NSF/NASA report (NSF/RA/N–73–001):

Conclusions

- Solar energy is received in sufficient quantity to make a major contribution to future U.S. heat and power requirements.

- There are numerous conversion methods by which solar energy can be utilized for heat and power; e.g., thermal, photosynthesis, bio-conversion, photovoltaics, winds and ocean temperature differences.
- There are no technical barriers to wide application of solar energy to meet U.S. needs.
- The technology of terrestrial solar energy conversion has been developed to its present limited extent through very modest government support and some private funding.
- For most applications the cost of converting solar energy to useful forms of energy is now higher than conventional sources, but because of the increasing prices of conventional fuels and the increasing constraints on their use, solar energy will become competitive in the near future.
- A substantial development program can achieve the necessary technical and economic objectives by the year 2020. Then solar energy could economically provide up to: (1) 35 percent of the total building heating and cooling load; (2) 30 percent of the nation's gaseous fuel; (3) 10 percent of its liquid fuel; and (4) 20 percent of its electric energy requirements.
- If solar development programs are successful, building heating could reach public use within 5 years, building cooling in 6 to 10 years, synthetic fuels from organic materials in 5 to 8 years and electricity production in 10 to 15 years.
- The large-scale use of solar energy would have a minimal effect on the environment.

Recommendations

- The federal government should take a leading role in developing a research and development program for the practical application of solar energy to the heat and power needs of the United States.
- The solar energy R&D program should provide for simultaneous effort toward three main objectives: (1) economical systems for heating and cooling of buildings; (2) economical system for producing and converting organic materials to liquid, solid and gaseous fuels or to energy directly; (3) economical systems for generating electricity.
- Research and development should proceed on various methods for accomplishing the above objectives and that programs with phased decision points be established for concept appraisal and choice of options at the appropriate times.
- Environmental, social and political consequences of solar energy utilization should be continually appraised and the results employed in development program planning.

The Eagle Head radio repeater near Bozeman, Montana, is operated by solar panels from Solar Power Corporation. (Courtesy Exxon)

Although many of these individual programs are moving forward, the pace is still slow. The big growth in solar energy will come after its economic feasibility has been proved to the satisfaction of industry and utilities. But economic feasibility is more than just low cost for certain items; it is also the assurance that this energy will become economical on a large scale. Not only must solar power be competitive with other methods, it must also be readily available in large quantities at a realistic price, in the event that other energy sources are no longer acceptable for economic, supply, environmental, safety or political reasons.

Solar energy offers mankind so many alternatives that we can no longer ignore it. Eventually it will find its way into all levels of our energy-consuming society. But it would be more effective if the transition to solar energy (as well as to other natural forms) were well organized.

Aside from being an unlimited supply of energy, solar power offers another great bonus. It does not add any heat load to the earth (except for the end product of consuming electricity). Unlike fossil fuel and atomic plants, solar energy systems would not produce any additional heat that had not already existed, in the form of sunshine. This means that if man were to double, triple or quadruple the amount of electrical power, the earth would not have to carry a huge additional heat load. When we consider the fact that in the year 2000, the generation of electricity may be six times the 1970 level, the heat factor becomes very important.

As to the objections that solar power is not reliable because of changing weather conditions, the answers are immediately available. First of all, if we used the orbiting solar satellite system, clouds would not affect the supply. With regard to solar farms, salt beds and other types of systems could store the energy for several days and in the areas where the solar farms would be located, there is little likelihood that the sun would be obscured for a long time. Besides, the main electrical power source would not be concentrated in a single area, so the sun would be available in other locations.

Some years ago R. Buckminster Fuller coined the phrase Spaceship Earth in an effort to dramatize the fact that this planet is finite in its resources and that we are confined to it as we hurtle through space. However, it might be added that this spaceship has an auxiliary external power source which, for all practical purposes, is infinite.

10

The Power of the Stars

For billions of years a mysterious process has fired the glittering lights in the night sky. We now call this process fusion. The power of the stars may truly be man's guiding light to unlimited energy.

Although fusion is as old as the universe itself, knowledge, even recognition, of this process is relatively new. In fact, it is one of the greatest triumphs of twentieth-century science. During the 1930s a great number of experiments were performed to substantiate the conclusions of theoretical physicists who had predicted that the splitting of the atom would yield a specific amount of energy in the form of radiation and heat. It was also theorized that the opposite was true: if the nuclei of two atoms were combined, or fused together, energy would be released—in fact, much more energy than in the case of fission. During some experiments in 1932, which involved artificially accelerating atomic nuclei, scientists duplicated the process of the sun (fusion) on a very tiny scale and generated the first man-made fusion reaction. Twenty years later those experiments bore bitter fruit when the world witnessed the holocaust of the first fusion (hydrogen) bomb explosion in the Pacific Ocean in 1952. This demonstrated the fantastic amount of energy that could be released when the nuclei of atoms are fused together.

The year 1952 also saw scientists begin a formal research program to study the fusion reaction as a source of energy, a process called CTR (controlled thermonuclear reaction). At first, this work fell mainly in the domain of government projects, under the control of the Atomic Energy Commission. The first ten years of fusion research were a painful learning experience. In 1957 the General Atomic Company joined in the research program and fusion research began to spread to other organizations (mainly universities). General Atomic developed a reactor called the Doublet, which we will discuss shortly. In their struggle with the elusive technical and theoretical problems of fusion, scientists learned some of the basics about a totally new field called plasma physics. Plasma, as we shall see, is the very heart of a fusion reactor.

Fusion research picked up in the 1960s. Certain theoretical predictions were verified in the laboratories. Some experiments, aimed at verifying the feasibility of controlled fusion, had very encouraging results. Also in the early sixties more organizations, including the University of Rochester, the Los Alamos Laboratory of the University of California and KMS Industries, became active in fusion research. Meanwhile, other countries, chief among them Russia, were showing a great deal of interest in this new power source. The scientific world was startled when in 1963 the Lebedev Institute in Moscow reported the successful yield of a small amount of neutrons (the product of a fusion reaction) from one of its fusion devices. But for the most part, solutions to the many technical problems seemed almost insurmountable and progress was very slow. Around 1968, however, a new experimental device was developed which gave the concept of this power source a boost and the scientific world new hope regarding the achievability of controlled fusion.

From its inception in the 1950s the U.S. fusion program has grown through a series of projects that were supported with about $20 to $30 million per year to a major energy program with a budget of $168 million in 1975. But still, after 24 years of work, this very promising technology is still in the laboratory stage.

If scientists could attain efficient controlled thermonuclear reactions, our energy problems would all but vanish. First-generation fusion reactors using presently planned fuels would generate four times as much energy when compared by fuel weight as could nuclear fission reactors using uranium. By weight, fusion fuel could supply about 10 million times more energy than could gasoline. What makes this fact so tantalizing is that the fuel used in fusion reactors is none other than seawater, or a certain portion of seawater called deuterium.

Since such huge sums are involved, scientists have coined a new expression—Q, which stands for 1 quintillion (10^{18}, or 1 followed by 18 zeros) BTUs. Right now, world energy consumption is less than 1 Q. If the world population were to grow to 10 billion, and all these people enjoyed present U.S. standards of energy consumption, they would use slightly less than 3 Q. One of the materials that may be required for fusion systems is lithium. The total world supply of this material is equivalent to over 21 million Q. The supply of the other material used as a fuel in fusion reactors, deuterium, is equivalent to 7.5 billion Q. So there is enough fuel to run fusion reactors for billions of years.

Fusion is the combining of two nuclei to form a single larger nucleus. As shown in Figure 79, the nucleus of a deuterium atom, which is a hydrogen atom with one neutron (an isotope of hydrogen), collides with the nucleus of tritium (hydrogen with two neutrons, another isotope of

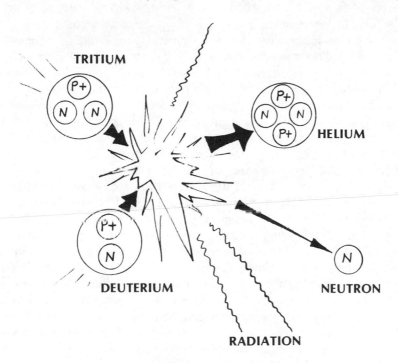

FIGURE 79. In a fusion reaction the nuclei of atoms are fused together to form larger nuclei.

hydrogen). If they collide with enough force, they will combine and form an entirely new atom (the forming of a new atom is called transmutation). However, the new atom does not weigh as much as the two original atoms;

when the atoms collided, one of the neutrons from one atom was broken loose and was emitted at high speed. Furthermore, some of the mass of the atoms was converted into energy in the form of heat and other forms of radiation. So each time a fusion reaction takes place, energy is generated. Unfortunately, the fusion reaction is not as easy to obtain as is the fission reaction. In the fission process all that is needed is to arrange enough fissionable material (uranium) in a pile and it will go critical. In fact, the problem is to keep the fission process from getting out of control. In fusion there is a totally different situation. The nuclei of all atoms contain the same type of charge; that is, each nucleus contains at least one proton, which carries a positive charge. But like charges always repel each other. Therefore, if two nuclei are to fuse, they must slam into each other with enough force to overcome the repelling force of the positive charge.

In order to accomplish this feat scientists are building devices to produce the conditions needed for fusion. This involves working with tremendous pressures and extremely high temperature. First of all, the scientists have got to get enough of the right type of atoms together so the reaction can take place. This is done by producing a very hot gas called plasma. Plasma is a collection of atoms that have been stripped of their electrons; so, in effect, it is a gas made up of atomic nuclei and free electrons, all going their separate ways. (It is also referred to as a completely ionized gas.) How does the plasma start a fusion reaction?

We are calling the plasma a fuel, so we have got to ignite the fuel to get the process started. Ignition here is the process of getting the nuclei to slam into each other with enough force to combine. In order for this to happen, the plasma must be compressed to a great density so that the atoms are closely packed. Also, the plasma must be heated to the unbelievable temperature of 100 million degrees c. This raises a number of questions. How does one go about compressing such a plasma? How can it be heated to such a high temperature? And once we've got it that hot, how do we contain and control the superhot plasma? Slowly, some answers are coming. First of all, one might take a look at the sun, where fusion is taking place in a big way. This huge ball of blazing gases is held together by the force of gravity. Perhaps we could discover a way to simulate this force. The closest thing to gravity is a magnetic field, and this is what scientists are using to contain plasma.

Figure 80 shows the basic setup. Plasma is introduced into a special chamber, which also supports a number of very strong magnets. The magnets establish a magnetic field, which surrounds the plasma and forces it to remain suspended in the center of the chamber. The magnetic field is

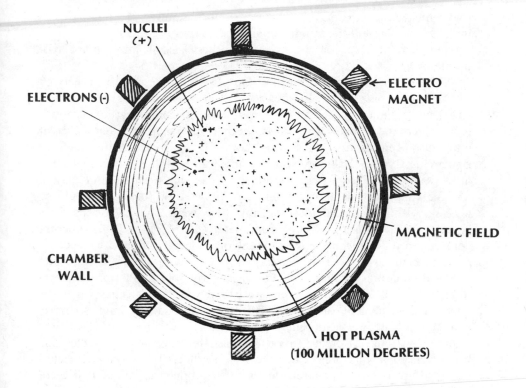

FIGURE 80. The tremendously hot plasma in a fusion reaction is suspended by strong magnetic fields.

increased in strength and, as a result, the plasma is pushed and squeezed into a smaller space. As this occurs, the ions (nuclei of the atoms) begin to collide more frequently, creating heat, so the temperature of the plasma rises. As the strength of the magnetic fields continues to increase, the plasma is further compressed and therefore its density (the number of atoms in a given space) also increases; so does its temperature. If it were possible to increase the strength of the magnetic field sufficiently, one could increase both the density and the temperature of the plasma to the point where ignition would occur and the fusion reaction would result. Unfortunately, this is easier said than done. Aside from attaining a temperature of 100 million degrees and an extremely high density, the plasma must be held in place for a certain length of time. This time varies with the density

—the higher the density, the shorter the time required. In other words, there has to be sufficient time to allow the fusion process to generate enough energy to keep the process going. Physicists ran into a number of problems.

The magnetic field did not raise the plasma temperature high enough for fusion to take place. Furthermore, it was found that plasma acts strangely; it is not very stable. Before the scientists could heat it very much or compress it, the plasma would leak out of the "magnetic bottle," as it was called. Temperature, density and duration—all three had to be of the right value. The scientists were able to achieve one value, but not all three at the same time. Years were consumed in studying the nature of plasma and the attainment of plasma stability. In the case of density, it would be necessary to squeeze the plasma until there were 100 trillion particles (nuclei and electrons) in 1 cubic centimeter ($10^{14}/C^3$), and this superhot compressed plasma would have to be held in place for about one second. Slowly, scientists around the world began reporting results. By the late 1960s the plasma stability problem was nearly solved. Plasma temperatures of 20 and 30 million degrees were achieved and densities (at lower temperatures) of 10^{14} and even $10^{16}/C^3$ were reached. But scientists could not put all three achievements together. The attainment of all three is required not only to create fusion but also to reach what is called the breakeven point, that point when the fusion process is well enough established to not only sustain itself, but also to produce an amount of extra energy that equals the amount of energy used to get the fusion process started. Obviously, in an electrical power generating plant the fusion process would have to produce energy considerably beyond the breakeven point. In fact, if a power plant were operating at about 30 percent efficiency, the fusion process would have to generate approximately 50 times more energy than the breakeven point to be economical.

In 1968 the picture became a lot brighter for scientists as a result of a Russian invention.

THE FUSION DOUGHNUT AND OTHER SHAPES

The breakeven point required the simultaneous attainment of three values: high temperature, high density and a long enough duration of plasma confinement. But in the early years scientists could not even approach these conditions; either the plasma would shift around, or it would leak out of the magnetic bottle or some other unexpected problem would occur. But at least they knew the target they were aiming for thanks to J. D. Lawson, a British physicist who in 1957 derived the basic values that defined breakeven. These values came to be known as the Lawson criterion, which is still used by scientists as a convenient standard to measure the extent to which plasma energy loses must be controlled so that a practical fusion reactor can be constructed. Aside from experimenting on ways to keep the plasma stable, scientists were also searching for new ways to superheat this strange gas. To do these experiments a large number of different fusion machines were built which basically fall into two categories: closed and open systems (see Figure 81). In the closed system the plasma is contained within a magnetic field (magnetic bottle), which is shaped like a doughnut. The plasma closes a loop on itself; hence the name, closed loop. The open system, also called a linear system, takes the shape of a straight tube slightly compressed near each end. In this system there is an open escape path for the plasma at each end, which has caused the problem of plasma escape. Open-ended systems are usually not suitable for plasma stability except when special devices are used, called Joffe bars after the Russian inventor of that name. The Joffe bars add a special type of magnetic field around the plasma to keep it stable. But even under the best conditions scientists were not getting very far with their fusion ma-

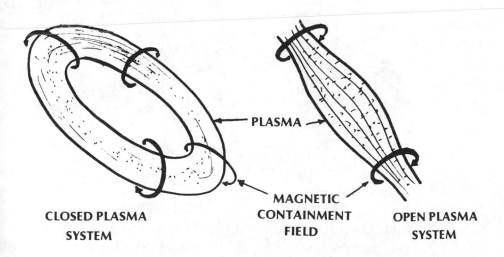

CLOSED PLASMA
SYSTEM

PLASMA

MAGNETIC
CONTAINMENT
FIELD

OPEN PLASMA
SYSTEM

FIGURE 81. All magnetic fusion machines can be reduced to two systems.

chines. Then came Tokamak, and a lot of things changed for the better.

Around 1968 the Russians announced the successful development and operation of a new fusion machine that solved a lot of the basic problems. Although it was not designed to demonstrate controlled fusion in the full sense of the term, the Tokamak achieved plasma stability, longer containment duration, higher density and higher plasma temperatures. The scientific world was taken by surprise, and at first was rather skeptical. But after a number of verifications were made, everyone became a believer in the Tokamak machine. Tokamak is a closed-system, doughnut-shaped fusion device, but with extra features. The Tokamak had been developed at the I. V. Kurchatov Institute of Atomic Energy in Moscow under the brilliant physicist Lev Artsimovich. After years of work it had been realized that it is extremely difficult to raise the temperature of plasma to ignition level merely by compressing it with a magnetic field. Also, scien-

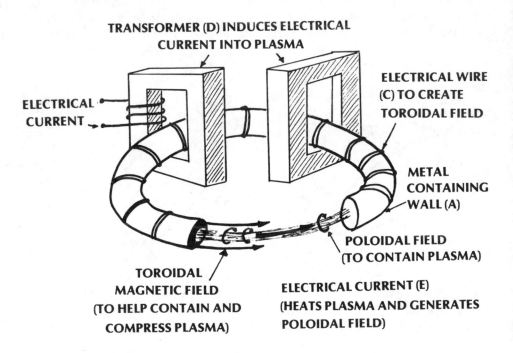

FIGURE 82. Russian Tokamak fusion machine.

tists found that to stabilize plasma, a number of magnetic fields oriented in two directions would be needed. The Tokamak solved both problems. As shown in Figure 82, this device has a metal tube (A) shaped like a dough-nut in which is contained the plasma (B). First, there is a set of electrical wires (C) forming a coil around the metal housing. This coil produces a magnetic field called a toroidal field, which is parallel to the shape of the plasma. This toroidal magnetic field helps to contain and compress the plasma. Another device called a transformer (D) is used to induce an electrical current (E), which flows right through the plasma. This electrical current does two things: it produces another magnetic field called a poloidal field (F) around itself, which also helps to contain the electrical current, and it generates a great amount of heat as it flows through the plasma thereby raising its temperature (this is called ohmic heating). This is where the name Tokamak comes from; *tok* in Russian means current. With this machine the Russians reached a plasma temperature of 20 million degrees and a density of 5×10^{13} particles per cubic centimeter, and held the heated plasma for about 30 milliseconds (20-thousandths of a second). This accomplishment was beginning to approach the Lawson criterion.

As a result of this success, many similar fusion machines were built in the United States, Great Britain and France. The first Western Tokamak was operational by May 1970 at Princeton University. Actually, the Princeton fusion machine, called a Stellarator, was very similar to the Tokamak, so scientists were able to make a quick conversion. But as good as it was, the Tokamak had some limitations. It seems that when plasma is heated by the electrical current and magnetic compression, its natural resis-tance to electrical flow is reduced. Reduced resistance results in less heat generation. So the ohmic heating method has a natural limit. Scientists in the United States began to modify the basic Tokamak. At Princeton, under the leadership of the well-known atomic physicist Dr. Harold Furth, a variation of the Tokamak called the ATC (adiabatic toroidal compressor) was built. The metal (copper) housing around the plasma was removed and a different type of transformer was used (see Figure 82). The results were highly successful; plasma densities twice as high as the Tokamak's and much higher temperatures were achieved. These encouraging develop-ments have convinced the federal government to fund the construction of a huge Tokamak-type fusion machine (with the latest design features), which will cost about $215 million. Meanwhile, other fusion devices, some which are related in design to Tokamaks, are under construction. One such device called the Doublet, built by General Atomic, uses a plasma housing that is somewhat like a figure-8 instead of circular in shape; however, both are shaped like a doughnut (see Figure 83). The Doublet appears to offer some advantages in plasma density not attainable in the basic Tokamak. Other fusion devices are quite different from the Tokamak family.

Not all fusion physicists were working with doughnut-shaped machines; some were working with open-ended systems. One such device is called the theta pinch, which looks like a long pipe (see Figure 83). In this machine the plasma is contained in the center of the tube and is heated by a method called shock heating. This is accomplished by inducing a second magnetic field very rapidly, so rapidly that the plasma is shocked by the force and quickly rises to a high temperature. The scientists at Los Alamos Scientific Laboratory are using one of these types of machines called

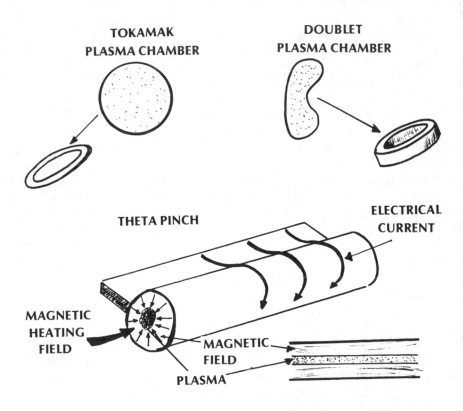

FIGURE 83. Different types of machines being used to solve the problems of controlled thermonuclear fusion.

Scientists will study plasma at a temperature of 200 million degrees in this newly constructed vacuum chamber at United Technologies Laboratory. (Courtesy United Technologies)

Scylla. This was one of the first U.S. machines to reach fusion ignition temperature.

One of the serious problems with open-ended systems is that the plasma has a tendency to travel along the lines of the magnetic field and escape from the ends of the magnetic bottle. After much research, however, scientists came with a new design called the magnetic mirror. In this device (Figure 84) the magnetic fields are formed by magnets that are shaped somewhat like the stitching on a baseball. In fact, two of these types of machines used at the Lawrence Radiation Laboratory at the University of California are descriptively called Baseball I and Baseball II. The magnetic field itself is shaped in such a way that the plasma is located on the lowest-strength level of the field, referred to as a magnetic well. The plasma cannot easily escape since whichever direction it tends to move in, it encounters a stronger field (the effect is as if it were trying to go uphill or climb out of a well). What happens in a magnetic mirror machine is that the plasma is reflected back and forth while it is being compressed and heated. However, this method cannot attain the Lawson criterion required for fusion. So the physicists, rather than try to adapt ohmic heating, introduced a new device called a neutral beam injector to the machine; here a supply of neutrons was generated and formed into a beam. These neutrons had been accelerated to a very high speed. Using a neutral beam injector (see Figure 84), the beam of high-speed neutrons is aimed directly at the plasma in the magnetic bottle. Since the neutrons have no electrical charge, they easily pass through the magnetic field and slam with great force into the plasma. This action causes the plasma to heat up rapidly to the fusion ignition point. Scientists hold out great hope for this method of heating. According to Dr. Melvin B. Gottlieb, head of Princeton's Plasma Physics Laboratory, neutral beam heating may lead to laboratory breakeven in the very near future. This type of plasma heating is being adapted to Princeton's new PLT machine (Princeton Large Tokamak). Meanwhile, at the Lawrence Livermore Laboratory, Dr. Frederick Coensgen is heading up an experiment called 2X IIB in which 12 huge neutral beams (shooting deuterium atoms) will be fired into the plasma being held in the magnetic bottle of its huge magnetic mirror fusion machine. Dr. Coensgen says that this new injection system will raise the plasma temperature to between 100 and 200 million degrees c. He goes on to state that the magnetic mirror machine has already had deuterium plasmas fusing to form the fusion reaction on a small scale. His group has also come close to attaining the right plasma density to keep fusion going. Dr. Coensgen is very optimistic about future developments. With larger funding he expects (if predictions

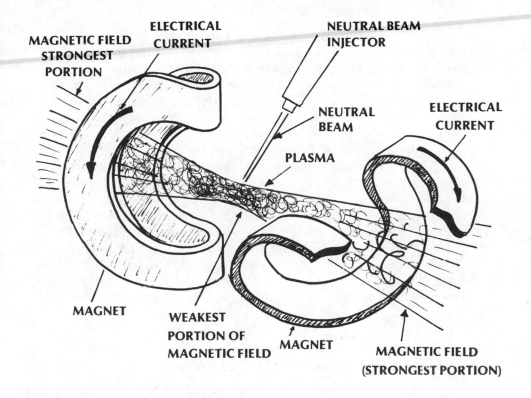

MAGNETIC FIELD STRONGEST PORTION

ELECTRICAL CURRENT

NEUTRAL BEAM INJECTOR

NEUTRAL BEAM

ELECTRICAL CURRENT

PLASMA

MAGNET

WEAKEST PORTION OF MAGNETIC FIELD

MAGNET

MAGNETIC FIELD (STRONGEST PORTION)

FIGURE 84. Magnetic mirror machine.

hold up and the plasma does not try any new tricks) to see a feasibility demonstration of controlled thermonuclear fusion in a very short while. That means that the fusion process in the machine will meet the Lawson criterion concerning temperature, density and containment time. This will lead to laboratory breakeven of plasma—in other words, the plasma will maintain the fusion reaction and will actually produce additional energy, which will be expelled from the ionized cloud (plasma). But this is only the very first step toward commercial fusion power.

What must be attained is nothing less than what is called system breakeven. Whereas plasma breakeven means that as much energy must be generated by the burning plasma as was used to ignite it, system breakeven means the burning plasma must produce excess energy equal to the energy that was used by the entire machine to ignite the system. This would include all the energy used by the high-power magnets and the ohmic heating current (if used) as well as the huge amount of energy that went into producing the neutral beams. That objective is still a decade away, though it could be speeded up with proper support from the government, industry and the power utilities. Already the scientists at the Lawrence Livermore have designed a 500-MW prototype fusion reactor power plant based on the magnetic mirror concept. This plant would use the heat from the reactor to generate steam, and it would also directly convert the charged particles from the fusion plasma into electricity.

The first-generation fusion power stations will most likely be fueled by a combination of deuterium (heavy water) and tritium (an isotope of hydrogen). This combination requires the least amount of effort to attain controlled fusion. When fusion does occur, its energy will be used in the system illustrated in Figure 85 (a simplified diagram). A device called an injector will introduce deuterium and tritium fuel into the inner reactor chamber, which will be kept in a vacuum state (particles of air would inhibit the fusion reaction). The fuel would then be ignited (the neutral beam injectors are not shown in the diagram). Surrounding the inner

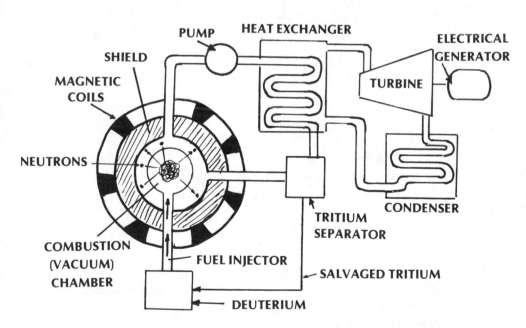

FIGURE 85. Thermal (heat) conversion type of fusion electrical power plant.

chamber would be a blanket of liquid lithium, which can absorb a great deal of heat. The lithium would be circulated by a pump through a heat exchanger. Here, the heat from the liquid lithium would be transferred to another medium such as water (or helium). The water would be changed into high-temperature, high-pressure steam, which would be used to drive a steam turbine.

After the lithium has passed through the heat exchanger, it would be sent into a special device called a tritium separator. One of the things that happens to the lithium as it surrounds the fusion chamber is that a small amount of it is changed into tritium. So since tritium is one of the fuels needed for the fusion reaction, it would be separated from the lithium and channeled into the fuel injector, which would inject it together with dueterium back into the fusion chamber. In the future such plants will be capable of producing 2 billion watts of electrical power.

But meanwhile scientists and engineers will have to develop the many special materials and techniques needed to construct and operate a full-size commercial fusion plant. Very powerful superconducting magnets must be designed. For shock heating, giant banks of electrical capacitors that can store the enormous amounts of electrical energy needed to rapidly increase strong magnetic fields will have to be developed. Further studies of plasma instabilities must be completed so that plasma reactions can be predicted accurately. Right now, no one knows what material can be used for the container walls and for the structure that will be bombarded by trillions of atomic particles. Other challenges are to determine how to inject the fuel so that it won't burn up before it reaches the center of the chamber, methods to ensure a high vacuum in the chamber and a host of other technical developments. New techniques now being perfected will allow laser beams to be injected into the plasma chamber to measure such things as energy level, density and temperature. Finally, a number of studies must be completed to determine if a laboratory-sized fusion reactor can be scaled up (made much larger) to the size of a giant commercial power plant without creating serious unforeseen technical problems. After all, the fact that a process works well on a very small scale is no guarantee that it will be practical or even work at all on a very large scale. Some basic technical changes may have to be made when engineers begin to construct large fusion plants. Some of the problems that face physicists are related to magnetic containment-type fusion machines. It is possible, however, that many of these problems can be circumvented by a different approach to controlled fusion.

The giant fusion machine called Baseball II is being worked on by scientists at the Lawrence Livermore Laboratory. (Courtesy Lawrence Livermore Radiation Laboratory)

LASER FUSION

So far, all the methods discussed and their associated problems have involved magnetic containment machines—Tokamak, theta pinch, magnetic mirror and others. However, there is a very different approach to fusion power that may very well circumvent the large number of problems associated with magnetic confinement methods; this approach is called laser-induced fusion. After all the attention given to other fusion methods, the laser approach could turn out to be the dark horse.

The laser (which stands for Light Amplification by Stimulated Emission of Radiation) was invented in 1960 simultaneously by American and Russian scientists—in fact, each team of scientists won a Nobel prize. In simple terms, a laser is a device that is capable of producing a highly concentrated beam of very special light energy. The basic operation of a laser is shown in Figure 86. A cylinderlike container holds laser material, which can be a solid, a liquid or a gas. A powerful light source is then concentrated on the laser material; this is called laser pumping. Some of the electrons orbiting the nuclei of the material will absorb some of this light energy (an action similar to that which takes place in solar cells). As shown in Figure 86(A), an electron that absorbs light energy will become highly energetic and have a tendency to move away from its position of rotation (1) outward to a farther orbit (2), and will remain in a high-energy state for as long as the light remains illuminated. When the light source is turned off, the high-energy electrons give up or emit the energy they had absorbed and return to their original orbiting level. This action takes place in many of the atoms of the laser material.

As shown in Figure 86(B), this emitted light travels in all directions, but some of the beams strike the highly polished mirrors at the ends of the laser cylinder, which bounce the light back and forth. On one side the mirror is only partially silvered and so some of the laser light is allowed to escape; this forms the well-known laser beam. However, the beam may not be very powerful as it exits, so a lens is placed in the path of the beam to intensify the laser energy as it strikes its target—and that target could very well be the plasma for a fusion reaction. Soon after the invention of the laser scientists got the idea of using this device to trigger a hydrogen bomb, which would be a "clean bomb" with far less radioactive fallout than the atomic bomb. As it has turned out so far, such an idea is very impractical. The laser system needed to trigger a hydrogen bomb would be too heavy to be carried on present-day aircraft (including the C-5-A). However, another area of interest to the military is laser methods of simulating H-bomb

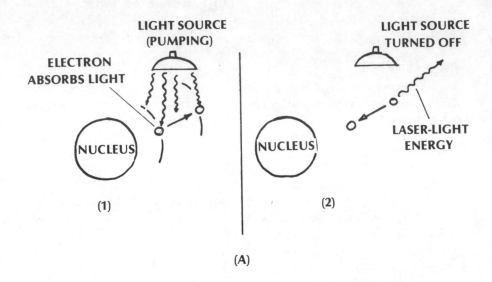

LIGHT SOURCE
(PUMPING)

ELECTRON
ABSORBS LIGHT

NUCLEUS

(1)

LIGHT SOURCE
TURNED OFF

NUCLEUS

LASER-LIGHT
ENERGY

(2)

(A)

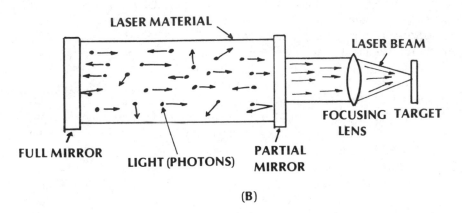

LASER MATERIAL

LASER BEAM

FULL MIRROR

LIGHT (PHOTONS)

PARTIAL
MIRROR

FOCUSING TARGET
LENS

(B)

FIGURE 86. A laser produces a highly concentrated beam of special light.

explosions. As we shall see, laser fusion involves the detonation of miniature hydrogen bombs, along with the atomic particles and X-ray radiation
that accompanies such an explosion. The result of this military interest was
that practically all laser research performed by the government during the
1960s was highly classified until 1971. Meanwhile, the Russians had been
carrying out similar research, and to a certain extent a race to fusion (not
as intense as the space race) has developed between the two nations. The

question is: Who will achieve laboratory breakeven first? Surprisingly, there has been a great amount of cooperation and exchange of information between the scientists of both countries. For a number of years some laser experts were working with the idea of firing a powerful laser beam at a small target, a sphere-shaped droplet of deuterium and tritium, but they had no success. Such a method required a huge laser, far beyond the capabilities of any present laser or even any laser in the foreseeable future. But with the declassification of government-sponsored research, an entirely different approach to laser fusion was revealed. How this system works is shown in Figure 87. A powerful laser is used to fire a light beam into a device called a beam splitter. This unit divides the laser beam into two equal parts. Keep in mind that since both new beams come from the same source, they will remain perfectly synchronized in their movements. Both beams are then deflected at right angles by a series of mirrors and projected into laser-beam amplifiers, which increase the power of both beams. These amplifiers are on opposite sides of the fusion chamber. As each of the two beams leaves the amplifier, it is once again divided into two equal parts by a beam splitter, which results in the creation of four separate laser beams, each perfectly synchronized with the others (since they all originally came from the same laser source). Meanwhile, a small spherical pellet of fusion fuel (deuterium and tritium) is injected into the chamber. This pellet is suddenly hit on all four sides by the laser beams. The shock is so great that the pellet is heated to 100 million degrees in about one-billionth of a second. Actually, this is not a continuous process; that is, the laser beam does not fire continuously. Rather, it fires in a short pulse, ideally for about one-nanosecond (one-billionth of a second). When the fusion energy in the chamber has been dissipated, another pellet is injected and the laser fires once again. This is somewhat like a shooting gallery, where the rifle is the laser and the moving target is the fuel pellet. One appreciates the accuracy requirements of the laser once one understands that the fuel pellet is much smaller than the head of a pin, and this extremely small target is falling through space when the laser beams are required to converge on it from all sides. In a fusion power plant 12 or more beams will have to strike the target simultaneously.

In order to understand the idea of laser fusion, one must become familiar with what takes place in the fuel pellet when it is struck by the laser beams. It will be recalled that early attempts to ignite fuel pellets with laser beams were unsuccessful. This is because the energy could not be delivered to the pellet fast enough. As the fuel pellet received energy from the laser beam, it began to vaporize rapidly and expand into a hot cloud of

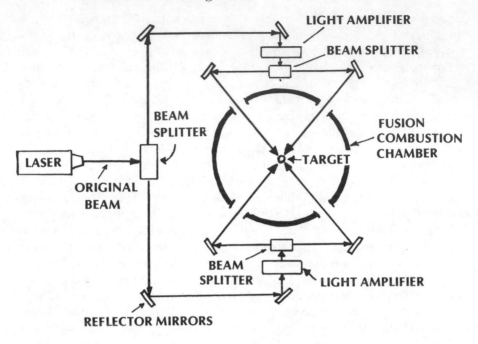

FIGURE 87. Laser energy causes a fuel pellet (target) to implode and ignite fusion process.

gas. As it expanded, it dissipated its heat and never approached the proper conditions for fusion (the Lawson criterion). In order to achieve fusion ignition with a single beam, scientists calculated they needed a laser capable of delivering 1 billion times the power output of the Grand Coulee Dam, or a laser about 10 million times more powerful than the most energetic laser then available. Such an idea was possible only in theory. But with the multibeam approach, conditions are quite different; here the pellet undergoes an implosion (the opposite of explosion). As was explained previously, in a laser fusion (also called inertial confinement) system a fuel pellet is fired into a chamber, at which time it is converged upon by a number of laser beams from all sides (see Figure 88). The pellet itself is extremely small, measuring only 500-millionths of a meter (about 14 of these pellets could be lined up across the head of a pin). As it is introduced into the fusion vacuum chamber, the pellet develops a thin atmosphere which is actually a boiloff of its surface material. The outer edge of this atmosphere is called the critical surface. It is this surface that

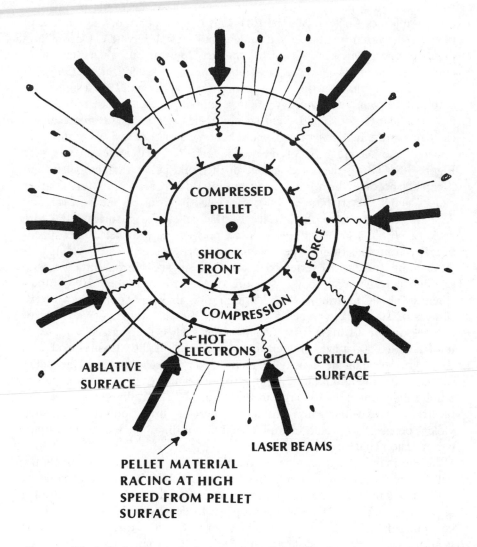

FIGURE 88. A fuel pellet must be compressed 10,000 times to attain fusion ignition.

receives the first impact of the laser beam energy, which would be possibly 1 million joules (1 joule is equal to 1 watt for 1 second, or the amount of power that would illuminate a flashlight bulb for 1 second). One million joules of energy may not sound like a lot of power, but when that entire amount of energy is delivered in one-billionth of a second, it is equivalent to delivering 1 trillion watts of power to that small target. This results in an extremely violent reaction. Two things happen in the pellet atmosphere: first, the heat is distributed evenly around the pellet; second, many of the electrons in the fuel pellet atmosphere become superhot and carry this heat down to the second level of the pellet, called the ablative surface (see Figure 88). This is the outer surface of the solid portion of the pellet, and it rapidly becomes heated to many millions of degrees. This temperature causes pellet material to be burned, violently torn off the surface and accelerated away from the surface of the pellet like so many miniature rockets. Within a few billionths of a second these tiny rockets reach speeds of many thousands of miles per second. In reaction to this accellerating pellet material, forces are exerted back onto the surface—in fact, the fantastic amount of force equal to the pressure of at least 1 trillion (10^{12}) atmospheres (about 15 trillion PSI). Similar pressures are found only in the centers of stars that are being compressed by the weight of millions of trillions of tons of mass. In a fusion pellet such a pressure compresses it until it is 10,000 times greater than its normal liquid density, and 100 times greater than the density of lead. Under these conditions the pellet reaches fusion ignition and a minihydrogen bomb explosion occurs, generating a huge amount of energy in the form of heat, radiation and atomic particles (mostly neutrons). In a commercial fusion power plant such violent occurrences would take place about ten times per second in a number of fusion chambers.

One may begin to appreciate the problem of energy requirements for pellet implosion by considering the fact that although a laser capable of delivering 1 million joules of energy in one nanosecond (one-billionth of a second) is required, plans are in progress to develop a 10,000-joule laser system. Although this represents only 1 percent of the goal, it will be a few years before even this system is perfected. But some progress is being made; the development of many types of laser such as neodymium-glass, carbon-dioxide, iodine-xenon and hydrogen-fluoride will offer a number of avenues to approach the ultimate goal. Aside from the laser, a lot of other technical problems must be considered.

For example, in order to compress a pellet 10,000 times, it must be spherical to a very accurate degree. How accurate? Scientists are still trying to determine. If it is not spherical enough, the pellet will disintegrate and dissipate its energy prematurely and fail to attain ignition. In order to determine the best possible shape and makeup of a fusion fuel pellet, scientists such as John L. Emmett, John Nuckolls and Lowell Wood of the AEC's Lawrence Livermore Laboratory are putting the world's most powerful computer (CDC 7600) to work for them. The highly complex program, called LASNEX, measures 60 values (variables) at 2000 different points in space and 10,000 moments in time. This results in the generation of a billion numbers. Such extensive programs will lead to the design of a high-performance fuel pellet.

But how does one go about manufacturing millions of such pellets, which must cost less than one cent each, and maintain a nearly perfect spherical shape? Then there is the combustion chamber; it must be maintained in a high-vacuum state (impurities would inhibit the fusion process) under high-temperature conditions. There is a lot more to be learned in that area. And the structure of the chamber itself—what material can be used to withstand possibly as many as 100 mini-H-bomb explosions per second? Further, the vacuum chamber walls (referred to as the first wall) will have to withstand intense radiation and bombardment by atomic particles (mostly neutrons) over a period of 10 to 20 years.

And this would be a pretty tough 20 years for the fusion chamber or first wall (see Figure 89). The actual fusion burn period will be less than a billionth of a second, and during this fantastically short period of time the fireball within the chamber will generate the equivalent of 1 quintillion (10^{18}) watts of thermal power. This is greater than the power used by all man-made machinery combined and ten times more than the total amount of sunshine that falls upon the earth. Such an atomic fireball also emits a shower of atomic debris. These high-speed particles go slamming against the chamber walls, as often as 100 times per second. Fortunately, all the particles do not travel at the same speed, so they do not hit the wall at the same time, but they all do some damage to the structure. So extensive will be the atomic particle bombardment that in the course of a single week's operation every atom in the vacuum wall will have been displaced at least once due to neutron collisions.

High-energy neutrons, for example, pass through the first wall (Figure 89) rather easily, but they leave scars in the form of dislocated and disintegrated metal atoms (vanadium). Some of the atomic collisions in the wall result in the formation of gas bubbles in the structure; the pressure of this atomic gas can rise to thousands of pounds per square inch and eventually rupture the reactor structure. Furthermore, many of the atomic par-

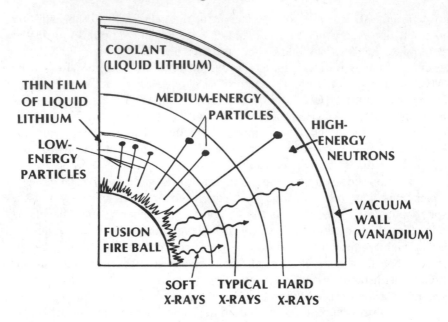

FIGURE 89. Fusion power plants will have to withstand almost continuous bombardment by X-rays and atomic particles.

ticles that lodge in the structure are radioactive, which further complicates the problem.

Another threat to the first wall of the fusion chamber comes in the form of X-rays that strike it and cause it to heat up. The difference in temperature between the inner and outer sides of the wall can lead to excessive stress and may cause the front wall to start flaking. One possible solution to this problem would be to construct the first wall out of a softer metal such as carbon, lithium or beryllium. In this way the X-rays would penetrate deeper into the wall and distribute their heat energy more evenly, thereby preventing a stress condition.

The chamber design itself must also be worked out in detail. For example, how does one prevent the atomic particles and the radiation from escaping from the laser and fuel entry ports leading into the chamber? One answer is illustrated in Figure 90. The fuel port will be shaped in an S-type bend, and the fuel pellet will be electrostatically guided around the bends and into the chamber. But the atomic particles and radiation would be traveling too fast to negotiate the turns and would therefore be blocked

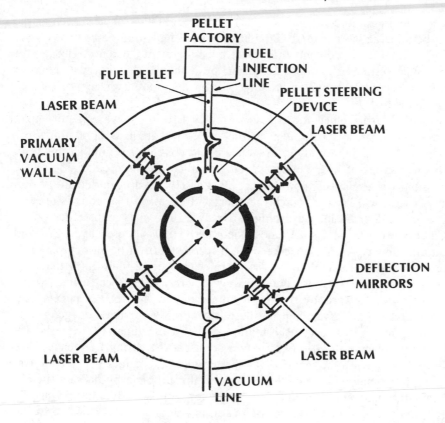

FIGURE 90. Fusion chamber design will prevent neutrons and radiation from escaping from the laser and fuel entry ports.

from escaping from the chambers. A similar method would be used for the laser beams. Only in this case mirrors would be used to bend the light beam and guide it into the chamber. The position of all the mirrors (96 or more) would have to be set with extreme accuracy to ensure that the beams converged at a precise point.

One of the most challenging problems will be system efficiency. As in the case of magnetic confinement, system breakeven means that all the energy required to produce laser ignition must be equaled by the energy generated by the fusion process. However, in the case of laser fusion the scientist is up against the fact that most lasers are extremely inefficient. Some exhibit efficiencies as low as .001 percent (i.e., they consume

100,000 times more energy than they produced). Other devices such as the carbondioxide lasers can attain efficiencies as high as 60 percent. But they suffer from other problems that affect overall system efficiency. Because of very low system efficiency, the fusion burn in the chamber will have to produce about 100 times more energy than the amount of energy used to ignite the fuel.

Aside from efficiency, a laser used in a fusion power reactor might have to fire its powerful beam at fuel pellets as often as 6000 times per minute. No laser presently in existence could possibly accomplish this task. But new types of lasers under development are possible candidates. One possibility is a relatively new device called an HF (hydrogen fluoride) laser. It is a chemical laser in which the energy of chemical reactrons is used to produce a laser beam. According to Dr. Reed Jensen, group leader of the Laser-3 program at Los Alamos, their new HF laser uses hydrogen and fluorine under pressure, together with a special electron beam pumping system (required to stimulate the laser) to release more than 2300 joules of laser energy in 30 nanoseconds. This will surely help laser fusion experiments move forward by delivering more energy to the fuel pellets. Dr. Jensen says that a number of these powerful lasers may be ganged to produce an energy level of 1 million joules. Unfortunately, there is one major problem with HF lasers; they operate at the wrong frequency—five times lower than that required for deuteriums-tritium fuel pellet compression. But it may be possible to get around this problem by passing the HF beam through a device called a frequency doubler. By multiplying the frequency a number of times, the beam may be suitable for use in fusion pellet compression.

Finally, there is the problem of optical integrity; all the mirror surfaces within the fusion reactor will be partially subject to the stresses of the fusion implosions. Their surfaces will have to be protected from X-rays and atomic particles that result from the implosions. All these and many other problems will have to be conquered before the first prototype fusion power plant is put into operation.

The generation of fusion activity is only part of what goes on in a power plant. Once fusion is accomplished, the energy from the process must be collected and converted into electrical or some other form of usable energy. In the first fusion reactor a thermal cycle will be used. About 80 percent of all the energy in a fusion reaction is carried off by high-energy neutrons. In fact, each neutron will be at an energy level of 14,000 electron-volts. When these neutrons slam into the lithium in the cooling blanket that surrounds the fusion chamber, they are stopped cold and all their kinetic energy is converted into heat energy, which is carried off by the liquid lithium to the heat exchanger, where it produces steam. Such a cycle will be about 60 percent efficient—which, while far from ideal, is considerably higher than the most efficient fossil-fuel or fission atomic plants (about 40 percent). With more advanced fusion plant approaches there will be other methods of extracting the energy.

Some types of fusion fuels produce mostly charged particles when they burn; these fuels will be discussed in a following section. But the existence of charge particles from the fusion fireball allows for more direct methods of energy conversion. One of these ways is MHD (magnetohydrodynamics). Since the fireball is composed of a superhot fluid (100 million degrees), under the right controlled conditions, this fluid could be exhausted through an MHD channel as a source of hot gas, and the hot gas could be used to generate electricity.

Direct conversion from the fireball to electrical wires is the object of still another fusion system. Here, the charged particles (positive nuclei and negative electrons) are captured and used as electricity. As shown in Figure 91, a dense stream of electrons, protons and alpha particles (positively charged nuclei) leave the fusion chamber guided by a strong magnetic field. According to Dr. Richard F. Post, inventor of this method, the particles in the stream are packed too closely together and the space-charge effect makes them very hard to steer. So the plasma is allowed to pass through an area that expands in size, causing the atomic particles to separate. Once this happens, the electrons can be extracted by a collector that has a positive charge on it; since the electrons are negatively charged, they are attracted to the collector, which guides them on to an electrical conductor (wire). Meanwhile, the more energetic and massive positively charged particles continue along the structure. The next object is to capture them. This is the job of another set of collector plates. However, at the speed these particles are traveling, if they were to strike one of the plates, all their energy would be dissipated as heat; so they have to be slowed down or decelerated. This is done by a series of plates that tend to oppose or repel the particles, thereby slowing them down. Once they are slowed to a reasonable speed, they can be collected by the collector rings and guided onto the electrical conductor wires. The result of all this activity is the generation of DC (direct current) electricity, which can then be converted to AC (alternating current) for home use. The system proposed by Dr. Post would involve a structure measuring about 500 feet in diameter, but it could result in a method of generating electricity approaching 90 percent efficiency.

Most likely, fusion power plants called hybrids will be developed. These reactors will use a combination of methods to convert the fireball

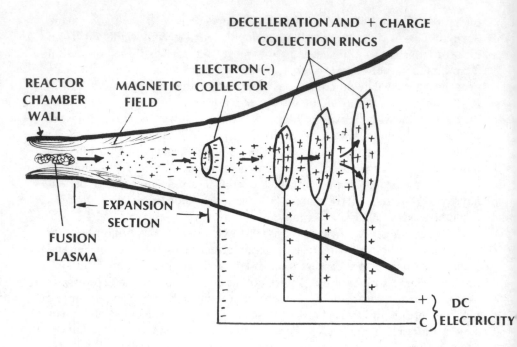

FIGURE 91. Direct conversion of hot fusion plasma into electricity.

energy into electricity. They will use both the thermal (heat) cycle and the direct power conversion methods to take advantage of all the reactor's energy.

A great deal of the technology of plant construction, fuel handling, energy conversion, reactor structure and so forth depends on what kind of fusion fuel (fuel cycle) is being used. A number of different possibilities exist and each fuel combination results in different levels of energy generation and atomic particle production. The first fusion power plants will almost certainly use a combination of deuterium and tritium because these offer the easiest path to fusion. For example, this combination has the lowest ignition temperature (100 million degrees). And as was mentioned previously, deuterium is extremely abundant and available at an insignificant cost when it is factored into the overall cost of fusion power. Although tritium is expensive and is not found in nature, it can easily be bred in the lithium cooling blanket that surrounds the fusion combustion chamber.

When a neutron strikes a lithium atom, one tritium atom and one helium atom are formed and 4.8 million electron-volts of energy are released in the process. Therefore in a fusion reactor it will only be necessary to supply tritium in the beginning (about a month's operating supply). After that, the reactor will generate its own tritium from the lithium. Happily, lithium is available in great abundance. At present prices the cost of fuel for a fusion reactor would be less than 1 percent that of coal. In fact, according to Tihiro Ohkawa, a director of Gulf General Atomic's fusion program, it costs only $150 per pound. This means a fuel cost of only 0.003 mill per kilowatt-hour—or, to put it another way, one pound of deuterium fuel has the equivalent energy of over 180,000 gallons of gasoline.

There are many other fuel combinations such as deuterium and helium, lithium 6 plus a proton, deuterium and lithium 6, and boron 11 plus a proton. In each of these the fireball generates mainly charged particles, which make possible the highly efficient direct energy conversion possible. And in the case of boron plus one proton, the fusion reactor would generate 1000 times less radioactive debris than a reactor employing deuterium-tritium.

THE GOOD AND THE BAD OF FUSION

Assume that a large fusion power plant has been constructed. What will be its advantages and disadvantages when compared to fossil fuel, nuclear fission (including breeder reactors) and other forms of electrical power generation?

One might start off by saying that with few exceptions, there are only good things to say about fusion power. When fully developed, it will be a virtually inexhaustible source of energy. It will be economical, clean and safe. Although it suffers from some radiation problems, they are miniscule compared to fission plants, and fusion technology, unlike fission, has the potential of eliminating almost all of its radiation problems. But let us take a look at the many aspects of fusion power in some detail.

As with any infant technology, when fusion reactors first become commercially available, their reliability will not be as high as that of the more mature power plant types. Areas where there will be relatively little experience include large special-metal structures, powerful superconducting magnets, potassium-steam turbines, generators, boilers and, to a lesser extent, large high-temperature vacuum systems. Because of this unfamiliarity, a certain amount of redundancy will be required, which can be eliminated as the technologies develop.

Fusion power plants, like other systems, will be vulnerable to operational failures. Clearly, care in design can eliminate many potential problems. But the number of potential problems will represent minimal hazards. Failure of a magnet would cause the plasma to strike the wall, extinguishing the reaction with relatively minor effects on the wall. Failure of a fuel injector would reduce the fuel supply, causing the plasma to slowly diffuse away. Failure of the on-site reprocessing system would result in an impure fuel replenishment, which would markedly reduce the plasma temperature and therefore the reaction rate. But as fusion engineers gain experience, they will improve the reactor design to a point where these occurrences will be minimized and plant dependability will be maximized.

The only possibility of radioactivity release during routine plant operation is tritium leakage. On the basis of preliminary design studies it appears that tritium leakage can be maintained at very low values. During routine power plant operation tritium is anticipated to be the only radioactive effluent, and it appears to be readily controllable. A tritium leakage rate to the atmosphere of 0.0001 percent per day (based on a system inventory of 6 kilograms of tritium) appears reasonable from a design standpoint. Assuming that this leakage is to be discharged from the reactor building through a 200-foot stack, the maximum concentration at ground level would be reduced to the point where it would give a maximum dose rate downwind of 1 milli rem per year i.e., less than 1 percent of the average dose to the population from natural radioactivity.

When considering radioactive waste, it should be noted that tritium is one of the least toxic materials and has the relatively short half-life (time it takes to decay 50 percent) of 12 years. And even this time is reduced to only three days since the tritium is usually recycled to the combustion chamber as fuel. Compare this situation with a nuclear fission breeder plant involving plutonium, which is one of the most toxic materials known and which has an active life of 250,000 years. In an article in *Science* magazine the eminent physicists R. F. Post and F. L. Ribe stated: "although fusion energy is a form of nuclear energy, it bears almost no similarity to 'conventional' nuclear energy, that is, energy from the fissile elements uranium, plutonium, and thorium. Compared to fission and its hazards—radioactive fission products, the potentially serious consequences of accidents due to loss of control or loss of coolant, and the problem of the proliferation of nuclear fission weapons—fusion can be made much safer."

According to Dr. William E. Cough, and Dr. Bernard J. Eastland, both well-established scientists in the field of nuclear physics, the biological-hazard potential of the tritium inventory in a deuterium-tritium reactor is less by a factor of a million than that of the volatile radioactive material iodine 131 contained in a fission reactor. What about long-term radioactive waste? In a fusion plant the problem is associated only with part of the reactor structure, but on an overall basis, the problem of long-term radioactive waste is extremely minor and inconsequential when compared to the same problems for the nuclear fission area, which by its very nature generates huge quantities of highly toxic, long-lived radioactive waste.

Fusion reactors will produce nonvolatile, long-lived radioactive wastes in modest quantities. The primary source of radioactive waste will be the activated structural material of the blanket, which will have a finite useful lifetime within the reactor owing to radiation damage.

The activated structure of a fusion reactor could be reused after reprocessing if there was a scarcity of niobium resources. In view of the rapidly growing use of automation in industry, the remote handling and recycling of radioactive material may prove practicable and economical, thus virtually eliminating the need for long-term radioactive waste management in a fusion power economy—the recycling of the blanket structure after allowing time for radioactive decay.

To start up a fusion power plant an initial fuel charge of deuterium and tritium will be needed. Thereafter a continuous supply only of deuterium and lithium will be required at the rate of about 1 kilogram per day. Tritium shipment will be necessary only to supply the initial charges to start new power plants—possibly about 10-kilogram quantities from each operating plant every few years on the average, depending upon the rate of growth of the fusion power industry. When the radioactive blanket structure is replaced, the used activated unit will have to be shipped from the power plant to a site where it would either be stored or reprocessed. The structure itself will be nonvolatile and consequently its hazard potential should be relatively low. It will not require a large amount of shielding during shipment nor present a difficult cooling problem.

Fusion reactors are very attractive when considered from the standpoint of accident potential. A runaway reaction will not be possible in a fusion reactor both because of the inherent nature of the plasmas and because of the low fuel inventory—about 1 gram—that would be resident in the core during operation. This is in sharp contract to fission reactors, which must contain a critical mass of fissionable material (enough fuel for many months of operation) and therefore an extremely large amount of potential nuclear energy. The stability and safety of such a large amount of atomic fuel is highly dependent on the continuous operation of a huge cooling system. In a fusion plant the fuel safety factor does not depend on the operation of the cooling system.

One concern is the possibility of the abrupt release of a substantial amount of energy via nuclear reactions. The only fuel that could possibly react would be that actually inside the plasma region—about 1 gram. If all of the fusion energy obtainable from this charge were to be dumped into the cooling blanket in a few seconds, the average temperature of the lithium would rise about 30° C.—a minor perturbation.

Another fear is that of the reactor's overheating because of the failure of the cooling system. In a nuclear fission reactor this would be a very serious situation, but in a fusion reactor things are quite different. Study of the after-heat problem indicates that it is possible to evolve a design that is virtually unaffected by a loss of coolant accident.

Fusion power plants can be designed with overall plant efficiencies greater than 50 percent. And there exist the possibility of eliminating almost all heat pollution in fusion reactors by using fuels that produce mainly charged particles. In this way a direct conversion method would be used, raising the efficiency extensively and drastically reducing excess heat. Therefore a fusion power plant will result in far less heat pollution than nuclear fission or fossil-fueled plants. Furthermore, the safety and environmental characteristics of fusion reactors should make them potentially acceptable for urban siting, which would further reduce total fusion power costs by savings in transmission costs as well as possible savings associated with the sale of waste heat for building heating and cooling and/or industrial processing. And if urban siting is indeed acceptable, then the large land areas usually required for power transmission from rural to urban areas would be significantly reduced.

The national security aspects of fusion power would be manifold. The United States has plentiful deuterium and lithium resources and would therefore be independent of foreign sources. Fusion reactors do not utilize fissionable materials, which may be subject to diversion for clandestine purposes. And there is no known way to construct a hydrogen bomb without using fissionable material (uranium, plutonium, etc.) to initiate or detonate the explosion. Therefore material diversion is a problem with fissionable material, not with tritium.

A mature fusion reactor industry would strengthen the country's technological base and foreign sales of fusion reactors would have a favorable effect on the balance of payments. Some reliance on foreign sources of materials such as nickel and chromium is inevitable.

One disadvantage of fusion plants is that they come in big packages; in order for them to be economical, they must be at least 1000 MW in size. Most likely, fusion plants will be built in the 2000-MW to 5000-MW range. But when it comes to merits, fusion wins the contest with fission or fossil fuel plants hands down. Whether judged on environmental effects, safety, economics, international stability or efficiency, fusion comes out way ahead.

So what is being done to make the dream of fusion a reality? What steps are being taken to solve the problems that need solving and to start building clean, safe fusion electrical power plants as soon as possible? Unfortunately, fusion has gotten a relatively slow start because most of the financial support has gone to other projects (nuclear fission has eaten up the lion's share). But a few private companies that believe strongly in the potentials of fusion power have been working quite hard and spending large sums of private capital on research. One of these is KMS, a small company in Ann Arbor, Michigan. It lays claim to the honor of being the first group to attain a fusion reaction yielding high-energy neutrons via the laser implosion method. It had been working on the project for a number of years and results were not too encouraging. Hundreds of tries had been made at firing a group of lasers at a fuel pellet target with no resulting neutrons, or not enough to be meaningful. Then at 4:30 P.M. on May 1, 1973, Ray Johnson, a young physicist supervising the experiments, removed the print of a Polaroid camera that had recorded the 1036th firing in the atomic shooting gallery. Johnson almost became hysterical. "We've got neutrons!" he shouted. A few minutes later Keith A. Brueckner, KMS's head theoretical physicist and inventor of the company's fusion approach, went out and returned with a case of champagne for a well-deserved celebration by the whole KMS crew. Things have been moving around the country a little faster since then.

Sandia Laboratories has recently reported the firing of a HF laser which has produced 4200 joules of energy with a pulse lasting 20-billionths of a second; this is equivalent to about 200 billion watts of power. The next step that will be taken in the fusion research area will be the design and construction of multibeam laser systems capable of concentrating 10,000 joules or 50 to 100 trillion watts of power on a fuel pellet in a fleeting period of one-half of a billionth of a second (500 pioseconds). With some modification, such a system should be capable of delivering as much as 50,000 joules of energy to the target in a few billionths of a second. A number of accelerated programs concerning various aspects of fusion are being conducted at locations all over the country including the giant laser project being worked on by Lawrence Livermore Laboratory and Sandia Laboratories.

Meanwhile, a lot of laser fusion work is being done by other countries. The Russians are planning to move directly into the construction of a 27-beam laser fusion machine, designated the T20. Work is also going on in France, England and Japan. Dr. Charles Gilbert has stated, "In laser-fusion research, Russia would have to be first, the United States second, and France third. Fourth might be either Japan or West Germany; I'm not sure which." One of the reasons the United States is second is that for many years funding for fusion programs has been far below what was required—certainly not as low as for solar, geothermal and other energy programs, but not high enough to have moved fusion along at a rapid pace. During the 1960s the government spent about $30 million per year on fusion; in 1972 it spent $39 million and in 1974 $58 million. Now ERDA (Energy Research and Development Administration) is pumping in $75 million for laser fusion in 1976, and will spend $80 million in 1977, $99 million in 1978, $114 million in 1979, $130 million in 1980 and $144 million in 1981. In all, ERDA will spend $11 billion on laser fusion between 1975 and 1995. One encouraging aspect of this funding is that much of the activity will be aimed at solving the engineering problems connected with fusion power. Furthermore, ERDA has announced plans to build a giant advanced fusion reactor designed to clear away the final technical problem in reaching a full fusion process (laboratory feasibility and plasma breakeven). If all goes well, the $215 million test reactor, to be constructed on the Forrestal campus of Princeton University, will go into operation in the early 1980s. It will be a giant step toward commercial fusion power.

When can we expect to have a commercial plan in operation? The time frame for such an accomplishment varies according to the persons being asked. The answers range from an optimistic 15 years to a dismal 50 years; most experts home in on the year 2000 as a reasonable date for the first commercial power plant to start pumping electricity into the nation's grid systems. According to Dr. William E. Shoupp, vice president in charge of research at Westinghouse Electric Corporation, there is no chance that controlled thermonuclear fusion will be ready before the turn of the century. He sees a fusion plant to demonstrate engineering in the 1990s. David Rose, professor of nuclear engineering at MIT, feels that another 25 years will be needed to iron out all the problems of fusion before we can move on to commercial power. On the other hand, John F. Clark, director of the Thermonuclear Division of Oak Ridge National Laboratory, thinks that fusion breakeven will occur around 1981. He also feels that the first big breakthrough will come, not with laser fusion, theta-punch or magnetic mirror machines, but with a huge Tokamak fusion machine. Even more positive is Dr. R. S. Cooper of Los Alamos Scientific Laboratory, who feels certain that a pilot laser fusion power plant could be operating by 1982, and that the United States could have commercial fusion power plants by 1990. What it all boils down to is effort—how great an effort are we willing to make to achieve this great energy prize? Many groups with commercial interests might be quite unhappy if federal funds were shifted away from nuclear fission to fusion research and development. But what is needed is just that very switch. In 1973 at an international conference on the world's energy crisis, a Swedish physicist, Hannes Alfven (one of the world's top plasma experts and a Nobel Prize winner), made a very bold statement in defense of fusion: "It is often claimed that fusion reactors will not be developed before the year 2000 and, therefore, fusion energy should not be mentioned in the present debate. The causality chain may be the reverse: as the breeder-reactor lobby does not like the competition with the fusion alternative, this is eliminated by the claim that it belongs to a very distant future."

The case for fusion has been stated very effectively by Dr. Keith Boyer of Los Alamos Scientific Laboratory: "Surely no other conceivable energy source, with the possible exception of solar energy, can come close to fusion in terms of fuel reserves, minimum hazard, minimum adverse environmental impact and potential low cost for producing electrical energy."

So there it is—the power of the stars at our fingertips. Certainly fusion, under the most intense program, will not begin to solve the energy problems of the 1970s or even the 1980s. But then, neither will the breeder reactor. However, what we as a nation are being asked to do is place our bets on nuclear fission to carry us for the next 25 to 35 years; it may carry us right to the grave. With fusion, we do not have to gamble with safety or international crises—it is an ideal energy source.

11

Making a Wish
Come True

The apparent energy crisis is a classic example of not seeing the forest for the trees. There is, in fact, so much energy available that at times it even becomes a threat to man. A fierce hurricane or windstorm will expend countless trillions of calories of energy in uprooting trees and leveling homes and property. A violent volcano will destroy a city in a terrifying demonstration of its mighty store of energy. The blazing sun will scorch the earth, laying waste thousands of square miles of land as it floods unbelievable amounts of energy upon our planet. Yet, in the midst of this, we stubbornly insist that there is an energy crisis. Why? Because of all the forms of energy available, we have chosen to base modern society on those fuels that are least available and nonrenewable, the fossil materials: oil, gas and coal. And as if matters were not bad enough, in seeking to rectify our error, we appear to have chosen an alternative form of energy that may well lead to an even worse predicament.

The energy crisis has many ramifications; some are felt directly, others are barely perceptible. Not only can we suffer on as individuals and as a society from a lack of energy, but we can also be adversely affected by our misuse of energy. Choosing the wrong types of energy sources has resulted in serious pollution; neglecting conservation and efficiency considerations has led to a depletion of fuel supplies, the creation of thermal pollution and a general waste of energy. But perhaps the most serious aspect of the energy crisis is our apparent lack of concern for the long-range effects of technological developments. Present-day automobile pollution is an excellent example of this lack of concern. In 1900 there were basically three types of cars under various stages of development—the steam-powered car, the electric car and a car that incorporated an internal combustion engine (which burns gasoline). The last car was developed and eventually sold by the tens of millions because it was the easiest of the three to manufacture and cheap fuel was available. However, if a group of scientists had been assembled around the turn of the century to ponder the situation and look ahead, they might well have recommended a different course of action. The electric car was very limited because of the limited power available in the batteries of that day. The steam-powered auto suffered from problems such as slow start-up and water freeze-up in winter. But it was an infant industry and these technical problems could have been overcome with time. Had the scientists of 1900 been able to persuade the car makers to develop the steam-powered car, had they informed the general public that large numbers of gasoline-burning automobiles could create air pollution problems, the pollution of our large cities would be far less serious today. And a lot fewer people would be dying from its effects. The National Academy of Sciences has reported that automobile exhausts alone cause 4,000 deaths each year and also result in 4 million days of illness annually. If the electric car had been chosen, cars would present no air pollution problem at all.

Today we are confronted with a similar situation. We have a choice of technologies which are being promoted by various individuals, groups and agencies. We have the opportunity to investigate and ponder the long-term results of each of these technologies. Which energy technologies are best suited to our long-range energy needs? We must choose energy technologies that will not only provide all societies with sufficient energy that will not pollute our environment and endanger our health or existence, but that will also minimize world tensions. This means technologies that will deny would-be terrorists or criminally ambitious nations a source of destructive power (nuclear weapons), technologies that will eliminate international competition for critically needed fuels (fossil fuels). With fossil fuels running out, nations that do not have the large funds or advanced technologies will need other sources of energy. They are being forced to turn to nuclear power because we have not taken the lead in developing alternative sources. Since we cannot offer them an alternative, we have become instrumental in their turning to atomic power. We must choose energy technologies that will enable all nations, big and small, prosperous and poor, to extract the energy they need to survive and develop at a reasonable level of human dignity. In short, we must choose energy technologies that will in no way threaten our physical or political tranquillity. Furthermore, the choice must be made soon; the energy crisis is steadily getting worse.

According to Frank G. Zarb, head of the Federal Energy Administration, "Everything is going the wrong way" in the United States' energy equation. "Production is declining, while demand is rising. We are now vulnerable to the extent that 38 percent of our supply [of oil] comes from abroad, and this soon will rise to 40 percent." This means that the United States has to pay out a whopping $25 billion abroad each year just for oil. Nor will we be able to ease this situation by developing much larger quantities of our own oil and gas reserves; they are fast running out. A recent geological survey reveals that we have far less undiscovered oil and gas than was previously thought. Estimated undiscovered oil reserves are now put at between 50 billion and 127 billion barrels, and natural gas liquids at 11 billion to 22 billion barrels. The total of 61 billion to 149 billion barrels is far less than even the low end of the previous estimate of 200 billion to 400 billion barrels. The natural gas situation is similar; studies indicate an undiscovered reserve of between 322 TCF and 655 TCF, sharply down from the previous estimate of 1000 TCF to 2000 TCF.

National response to our predicament must be extremely rapid. First, our entire attitude toward energy must change; each one of us must develop an awareness of the importance of energy in our lives. We must fully comprehend the fact that energy is more than a luxury, more than a practical convenience; it is necessary to our existence. Therefore our attitude toward energy should assume the same importance as our attitude toward our need for shelter, food and even the air we breath. Such an attitude should have a marked effect on every aspect of our lives—the way we drive our car, heat our homes, what we buy, what we discard, the way we design new appliances, new machines, new cars—for both conservation and the design of energy-efficient new products can result in a marked improvement in the energy picture. In other words, our nation must develop a strong energy awareness and an energy ethic. Only then will we be able to join in a coordinated national effort to solve the production half of the energy crisis.

A totally planned approach to electrical energy production is the key to our future. We should not be swayed by the cynics, pessimists, persons without vision or by groups and companies that are motivated by self-interest. If we set our minds to it, the United States could soon be drawing a significant portion of its electrical energy from natural sources. What is important is that these systems can work together in an integrated network. Wind power does not have to provide 100 percent or even 50 percent of our needs, nor does solar, geothermal, solid-waste or tidal power. But combined, they can go a long way in providing for our needs. For example, a combination of solar power and wind power works well as complementary power sources feeding an integrated network. When the sun is not shining, there is often a higher-level wind condition. Winter and summer also make such a combination complementary. Solar and wind power also complement each other when they are situated at locations separated by reasonable distances. With the development of large electrical networks or grids, made possible by superconducting transmission lines or even by microwave transmission, all the clean energy systems could be used to feed into the system, each one supplying a meaningful amount. But we should not judge that amount against national needs. A tidal power plant delivering 5000 MW of electricity might not be considered significant in terms of national needs, but in the upper regions of the Northeast it would represent a good supply of power. The same is true for wind power in the Northeast, Midwest, Texas, Northwest and the Great Lakes region. Solar power could be significant in the Southeast, Southwest and many other areas; the same is true of geothermal power. As for solid-waste systems, they could contribute energy in all parts of the United States.

There is no reason why, with an all-out national effort, these combined natural forces could not be supplying 5 percent of our electrical energy by 1981, 10 percent by 1985, 25 percent by 1990 and 50 percent by 2000. By the year 2010 all the natural clean energy systems, together with fusion power, could be supplying 75 percent or more of our electrical energy needs. With a real effort, by that year we could even generate 100 percent of our electrical needs with these systems. Of course, we should not view wind, solar, geothermal and sea thermal systems merely as means of generating electricity. They are also methods for extracting clean energy from nature—which can be used to produce liquid hydrogen and oxygen as fuels for automobiles and industry.

One of the reasons given by private industry and utilities for not developing the new technologies (some of them are a lot older than atomic power) is that they do not have sufficient funds. This could also have been said about atomic power, which required huge sums of money for research and development, yet private industry forged right ahead with hundreds of millions of dollars to produce marketable products. In the case of wind, solar, geothermal and other systems the risk capital needed might be quite high; certainly, more than any single utility could afford. But there is a way that this problem could be overcome, and utilities are best suited for this solution. Since most utilities do not compete with each other, they could form large area associations or even a national energy association similar to the Electric Power Research Institute. Each company, according to its size, would contribute a certain amount of money to the association, and under its auspices there would be established a national program dedicated to research and development of new forms of clean, safe energy including solar, wind, geothermal, fuel cells, MHD, solid waste and ocean thermal. Such a National Energy Association could have a research and development center to work on new electrical energy storage methods such as advanced batteries, flywheels, magnetic and chemical. The development of superconducting transmission and microwave transmission techniques could also be handled at the center. Such an association would greatly benefit all the utilities around the country, since it would have the financial power to take on huge projects that could never be handled by individual companies. Such a facility (or facilities) would undoubtedly apply for a number of patent rights, thereby earning additional income through licensing fees from manufacturers for products, materials or processes they had invented. This income would be used for further development and might also be instrumental in lowering the required contributions by member utilities. The utilities would benefit from the research and development through the availability of advanced, more efficient and cleaner energy systems. Furthermore, such an association would also have huge purchasing power for totally new energy systems. For example, if a power company in the Northeast wanted to produce a small network of five or ten windmills for local power generation, it would be difficult to convince a large manufacturer to tool up and establish a costly manufacturing facility for so small an order. On the other hand, a national utilities association could establish a standard design with maximum efficiency through research and development. Numerous members from different parts of the country could combine their order and purchase the systems through the association. Faced with an order of a thousand or more, a manufacturer would not find it difficult to justify establishing a new manufacturing capability. After all, it would involve not only the initial sale (which would be considerable), but also a servicing contract and the sale of replacement parts. So a National Energy Association could inspire entire new energy systems and markets, and they would be a protection for the private enterprise system. What it will take to establish such an association is a little mutual trust, some foresight and a healthy appreciation of the tremendous potential of the natural sources of clean energy. An alternative to establishing a new energy association would be to expand the membership of the present association, ERPI (Electric Power Research Institute), to include all utilities instead of just investor-owned utilities, and for ERPI to make a strong national commitment to develop environmentally attractive energy sources. What utilities must keep uppermost in mind is that they are not gas and electric companies, they are in the business of energy; therefore they must shoulder the responsibility of being leaders in the field of energy. If they fail to recognize this fact, then responsibility and the total control of all aspects of energy will ultimately fall into the hands of the federal government.

In the end it boils down to a few basic facts: we are rapidly running out of fuel; the fuel we depend on is causing international tensions and placing our nation in a highly vulnerable position; if we choose nuclear fission as a main source of energy, we will move into a series of even more devastating circumstances; electricity based on the abundant natural, clean, safe forces of nature readily available to us is the only sane approach; electricity as a basic source of energy means national independence. The energy is there in great quantities, ready for the taking. We have been wishing for it for millennia. Will we now make the needed commitment? Will we have the foresight and the courage to make that wish come true? If we do, we will never again have to return to the Electric Wishing Well.

Bibliography

A great deal of detailed information is available on the various aspects of energy generation and on the energy crisis in general. Unfortunately, most of the in-depth information is to be found in trade publications and scientific journals. There is, however, some excellent material available from the Government Printing Office. The following is a partial list of source material that contains further details on the energy categories covered in this book. Each entry has been coded to indicate degree of technical depth. (L) indicates a low-level technical article easily understandable by the layman. (M) indicates an information source that is partially technical; the lay reader may easily follow some parts of the material but may find other portions beyond his understanding. (H) designates material that requires a considerable technical background on the part of the reader. The reader should not be put off by the publishing dates because some of the basic articles in the various technologies were written during the early stages of their development.

GENERAL

(L) *The Futurist* (October 1973), entire issue.
(L) "Our Energy Supply and Its Future," *Battelle Research Outlook* (November 1, 1972), Vol. 4.
(L) "Nuclear Electric Power," *Science* (April 19, 1974), p. 351.
(L) "Low-Cost, Abundant Energy: Paradise Lost?," *Science* (April 19, 1974), p. 247.

(M) "Environment, Energy and the Need for New Technology," *EDN* (June 20, 1974), p. 53.

(L) "The Energy 'Joyride' Is Over," *Fortune* (September 1972), p. 99.

(L) *Energy: Technology for Self-Sufficiency*, Combustion Engineering, Inc., 1974.

(M) "A Systems Approach to Energy," *American Scientist* (July–August 1974), p. 438.

(L) "Sun, Sea, Wind, Geysers—New Energy from Old Sources," *U.S. News and World Report* (January 27, 1975), p. 37.

(M) "Energy Sources of the Future," *Heating, Piping and Air Conditioning* (May 1967), p. 107.

(M) "A Comeback for Reddy Kilowatt?," *IEEE Spectrum* (April 1972), p. 43.

(L) "Electromagnetic Flight," *Scientific American* (October 1973), p. 17.

(L) "Wanted, Superbatteries," *IEEE Spectrum* (July 1972), p. 42.

(M) "Two for the Road (Batteries)," *IEEE Spectrum* (July 1972), p. 43.

(M) "In Power, Longevity and Voltage, Batteries are Reaching New Peaks," *Electronic Design* (May 24, 1974), p. 28.

(L) "Electric Vehicles' Role in Mass Transportation," *Electric Vehicle News* (August 1972), p. 6.

(L) "The Electric Energy Economy," *The Electrical Distributor* (January 1974), p. 54.

(L) "Electricity Demand Growth and the Energy Crisis," *Science* (November 17, 1972), p. 703.

(L) "The Future for Electric Energy," *Public Utilities Fortnightly* (April 25, 1974), p. 41.

(L) "Energy Crisis: Surviving the Critical Years, 1975–1985," *Electric Light and Power* (October 1974), p. 7.

(M) "Energy Storage: Using Electricity More Efficiently," *Science* (May 17, 1974), p. 785.

(M) "Energy Storage: Developing Advanced Techniques," *Science* (May 24, 1974), p. 884.

(L) "Compressed Air Storage: The Answer to Your Peaking Problems?" *Electric Light and Power* (October 1974), p. E/G 6.

(L) "Flywheels," *Scientific American* (December 1973), p. 17.

(M) "Superconductivity: New Roles for an Old Discovery," *IEEE Spectrum* (December 1972), p. 53.

(L) "The Superconductors Are Coming," *Machine Design* (July 26, 1973), p. 76.

(M) "Superconductivity: Large-Scale Applications," *Science* (July 19, 1974), p. 211.

(L) *The Futurist* (February 1974), entire issue.

(L) "Petroleum Resources: How Much Oil and Where?," *Technology Review* (March/April 1975), p. 39.

(L) "Utilities: Weak Point in the Energy Future," *Business Week* (January 20, 1975).

(L) "Chemical Week Report: 3 MILL Power Where You Want It By the Year 2000," *Chemical Week* (March 15, 1969), p. 92.

(M) "Fuels of the Future," *Iron and Steel Engineering* (January 1974), p. 91.

(L) *Energy: Technology for Self-Sufficiency,* Brochure T15-4039, Combustion Engineering, 900 Long Ridge Road Stamford, CN 06902.

(L) "Energy and the Future," *A.G.A. Monthly* (January 1973), p. 25.

(L) "Energy: A Strategy of Diversity," *Technology Review* (June 1973), p. 26.

(L) "U.S. Energy Resources: Limits and Future Outlook," *American Scientist* (January–February 1974), p. 14.

(L) "An Integrated National Energy Research and Development Program," (April 19, 1974), p. 295.

(L) "Heat Pollution," *National Parks Magazine* (December 1969), p. 17.

(L) "Faisal and Oil," *Time* (January 6, 1975), p. 8.

(L) *Initiatives in Energy Conservation,* Staff Report Prepared for the Committee on Commerce, U.S. Senate, 1973, U.S. Government Printing Office: 98-002.

(H) "Some General Information on Air Pollution," *RCA Engineer* (September 1974), p. 42.

(M) "Energy Conservation—A 'Must,' " *Iron and Steel Engineering* (August 1974), p. 53.

(L) "The Energy–Environment–Economic Triangle," *Technology Review* (December 1973), p. 11.

(M) "Air and Water Pollution: The Inseparable Duo," *Actual Specifying Engineer* (June 1974), p. 72.

(L) "Coal Is Cheap, Hated, Abundant, Filthy, Needed," *Smithsonian* (February 1973), p. 23.

(L) "The West Falthmouth Saga—How an Oil Expert Twisted the Facts About a Landmark Oil Spill Study," *New Engineer* (May 1974), p. 32.

(L) "Acid Rain: Fossil Sulfur Returned to Earth," *Technology Review* (February 1974), p. 8.

(L) "Evolution of an Energy Crisis," *Iron and Steel Engineer* (July 1974), p. 65.

(L) "Power from Coal," Series in *Power* (February, March, April 1974).

(M) "The Gasification of Coal," *Scientific American* (March 1974), p. 19.

(L) "Capturing Clean Gas and Oil from Coal," *Fortune* (November 1973), p. 129.

(M) "K-T Koppers Commercially Proved Coal and Multiple Fuel Gasifier," *Iron and Steel Engineer* (August 1974), p. 33.

(L) "Energy Self-Sufficiency: An Economic Evaluation," *Technology Review* (May 1974), p. 23.

(M) "Shale Oil—Not Long Now," *Chemical Engineering* (May 13, 1974), p. 62.

(L) "Prognosis for Expanded U.S. Production of Crude Oil," *Science* (April 19, 1974), p. 331.

(L) "Timetable for Expanded Energy Availability," *Science* (April 19, 1974), p. 367.

(L) "Economic Strategy for Import-Export Controls on Energy Materials," *Science* (April 19, 1974), p. 316.

ATOMIC POWER

(L) "Is Nuclear Power a Panacea?" *Chemical Engineering* (April 1, 1974), p. 24.

(L) "Fission: The Pro's and Con's of Nuclear Power," *Science* (October 13, 1972), p. 147.

(L) "Atomic Power: A Bright Promise Fading?" *U.S. News and World Report* (June 10, 1974), p. 43.

(L) "Nuclear Power Plants: Boom or Blight?" *National Wildlife* (April–May 1971), p. 21.

(M) "Uranium Enrichment: Laser Methods Nearing Full-Scale Test," *Science* (August 16, 1974), p. 602.

(M) "The SNAP-19 Radiosotopic Thermoelectric Generator Experiment," *IEEE Transactions on Geoscience Electronics* (October 4, 1970), p. 255.

(L) "Nuclear Radiation, Insidious Polluter," *Popular Electronics* (February 1972), p. 26.

(L) "Plutonium: Questions of Health in a New Industry," *Science* (September 20, 1974), p. 1027.

(L) "Plutonium: Watching and Waiting for Adverse Effects," *Science* (September 27, 1974), p. 1140.

(L) "Six Advocates of Nuclear Power," *Technology Review* (October–November 1974), p. 79.

(L) "Industry Awaits Solutions to Problems of High-Level Radioactive-Waste Management," *Power* (December 1973), p. 23.

(M) "The Nuclear Fuel Cycle: What's Happening Today?" *Power* (September 1973), p. 29.

(L) "Radioactivity: Burden from a Boom," *Power* (October 1969), p. 66.

(L) "What's Holding up Nuclear Power?" *U.S. News and World Report* (November 26, 1973), p. 23.

(L) "Nuclear Waste Disposal in the Oceans," *Science* (September 27, 1974), p. 1183.

(L) "Nuclear Power: Hard Times and a Questioning Congress," *Science* (March 21, 1975), p. 1058.

(L) "The Proliferation of Nuclear Weapons," *Scientific American* (April 1975), p. 18.

(L) "100 Nations Grapple with a Nightmare: Uncontrolled A-Arms," *U.S. News and World Report* (May 12, 1975), p. 67.

(L) "Rush to Nuclear Power: A Chain Reaction in World," *U.S. News and World Report* (June 9, 1975), p. 64.

(L) "Disposing of Nuclear Plant Solid Waste," *Power* (November 1970), p. 78.

(L) "The Unsolved Problem of Nuclear Waste," *Technology Review* (March–April 1972), p. 15.

(L) "Nuclear Fuel Enrichment: Nobody Wants to Do It," *Electric Light and Power* (December 1974), p. E/G 3.

(L) "Rebirth of the Breeder," *Machine Design* (March 23, 1972), p. 20.

(L) "Breaking the Nuclear Habit," *The New York Times* (December 5, 1974).

(L) "Plutonium: A Hot Potato," *Technology Review* (January 1975), p. 50.

(L) "Mushrooming Spread of Nuclear Power," *Time* (September 9, 1974), p. 28.

(L) "Nuclear Safety: Critics Charge Conflict of Interest," *Science* (September 15, 1972), p. 970.

(L) "Nuclear Safety: Barriers to Communications," *Science* (September 22, 1972), p. 1080.

(M) *Congressional Record*, Vol. 118, Washington, D. C. (February 23, 1972), No. 25 (92nd Congress, 2nd sess.), p. 5.2405–9.

(L) "The Economics of Nuclear Power," *Technology Review* (February 1974), p. 14.

(L) "Floating Nuclear Power: Energy Wave of the Future," *Finance* (April 1974), p. 26.

(M) "Supplying Enriched Uranium," *Physics Today* (August 1973), p. 23.

(L) "The Death of All Children," *Esquire* (September 1969), p. 1a.

(L) "Citizens v. Atomic Power," *New Scientist* (November 23, 1972), p. 450.

(L) "The Big Blow-up over Nuclear Blowdown," *Fortune* (May 1973), p. 2.6.

Miscellaneous Power Systems

FUEL CELLS

(M) "Apollo Spurred Commercial Fuel Cell," *Aviation Week and Space Technology* (January 1, 1973), p. 56.

(M) "Are We Overlooking the Fuel Cell?" *Electrical World* (December 1970), p. 44.

(L) "Fuel Cells: Fact or Fiction," *Chemical Engineering* (May 27, 1974), p. 62.

(L) "Modular Fuel Cell Program Advances," *Aviation Week and Space Technology* (January 7, 1974), p. 54.

MHD

(M) "Magnetohydrodynamics: Can MHD Unlock Our Coal Reserves?" *Electric Light and Power* (September 1972), p. 60.

(L) "Lack of Financial Fuel Limits Progress in MHD Power," *Machine Design* (November 25, 1971), p. 34.

(M) "MHD Efficiency Gets a Charge," *Industrial Research* (November 1974), p. 21.

(M) "Missile Technology Applies to Utilities," *Aviation Week and Space Technology* (January 28, 1974), p. 56.

(M) "Magnetohydrodynamic Power: More Efficient Use of Coal," *Science* (October 27, 1972), p. 386.

(L) "Direct Energy Conversion," *Energy International* (January 1970), p. 10.

(M) "Air Force Revs up 58-MW MHD Project," *Electric Light and Power* (October 1974), p. E/G 21.

SOLID WASTE

(L) "Solid Future for Solid Refuse Disposal," *Power* (October 1972), p. 102.
(L) "Wattage from Waste," *Nation's Business* (June 1974), p. 44.
(L) "The Good from Garbage," *Time* (December 2, 1974).
(M) "Garbage Routes to Methane," *Chemical Engineering* (May 27, 1974), p. 58.
(M) "Closing the Refuse Power Cycle," *Combustion* (February 1974), p. 20.
(H) "Fluid-Bed Pyrolysis of Solid Waste Material," *Combustion* (February 1974), p. 13.
(L) "Better Hurry, the Garbage Is Going Fast," *Electric Light and Power* (June 23, 1975), p. 14.
(M) "Solid Waste and Sludge = Energy Self-Sufficiency," *Resource Recovery and Energy Review* (January–February 1975), p. 16.
(L) "Memphis Makes Its Waste Pay Off," *Business Week* (July 7, 1975), p. 60L.
(M) *Energy Potential from Organic Wastes: A Review of Quantities and Sources,* IC8549 Bureau of Mines Information Circular 1972, Washington, D.C., Government Printing Office.

HYDROGEN

(L) "Hydrogen: Transportable Storable Energy Medium," *Astronautics and Aeronautics* (August 1973), p. 38.
(L) "The Case for Hydrogen-Fueled Transport Aircraft," *Astronautics and Aeronautics* (May 1974), p. 40.
(L) "The Cleaning of America," *Astronautics and Aeronautics* (February 1972), p. 42.
(M) "Hydrogen, Simplest of Gases—New Storage System," *S/D* Canadian publication (1974/4), p. 32.
(L) "The Coming Hydrogen Economy," *Fortune* (November 1972), p. 138.
(H) "Hydrogen for Refining—Design and Performance," paper by R. N. Bery delivered at Symposium on Hydrogen Manufacturing, Chemistry and Catalytic Technology, Division of Petroleum Chemistry, Inc., American Chemical Society, Los Angeles Meeting, March 28–April 2, 1971.
(M) *Hydrogen Plants from Foster Wheeler,* brochure, Foster Wheeler, 110 South Orange Avenue, Livingston, New Jersey 07039.
(L) "Future Availability of Liquid Hydrogen," *Astronautics and Aeronautics* (May 1974), p. 55.
(M) "Nuclear Water Splitting and the Hydrogen Economy," *Power Engineering* (April 1974), p. 48.

(L) "Progress on Hydrogen as a Fuel," *Industrial Research* (March 1975), p. 22.
(M) "PSE&G Builds Innovative Hydrogen Storage Plant," *Electric Light and Power* (November 1974), p. E/G 22.
(L) "The Hydrogen Economy," *Scientific American* (January 1973), p. 13.

TIDAL POWER

(L) "The Call of the Running Tide," *Engineering Opportunities* (July 1969), p. 16.
(M) "A Formulation for Power: Mathematics of the Fundy Tides," *S/D*, Canadian publication (1974/4), p. 20.
(L) "The Quoddy Question—Time and Tide," *IEEE Spectrum* (September 1964), p. 96.
(L) "Ocean Wave Power Electricity—Producing Facility," *Machine Design* (February 1975).

WINDPOWER

(L) "Back to the Windmill to Generate Power," *Business Week* (May 11, 1974), p. 140.
(L) "Tilting at the Energy Crisis: A Windmill on Your Roof?" *Machine Design* (January 10, 1974), p. 20.
(L) "Can We Harness Pollution-Free Electric Power from Windmills?," *Popular Science* (November 1972), p. 70.
(L) "Windmills: The Resurrection of an Ancient Energy Technology," *Science* (June 7, 1974), p. 1055.
(L) "Wind Power: How New Technology Is Harnessing an Age-Old Energy Source," *Popular Science* (July 1974), p. 54.
(L) "Windmills in the History of Technology," *Technology Review* (March–April 1975), p. 8.
(M) "Wind Power's Contra-Rotating Generator," *Design News* (February 18, 1974), p. 66.
(H) P. C. Putnam, *Power from the Wind*, Van Nostrand, Reinhold Co., 1948.

GEOTHERMAL POWER

(L) *The Potential for Energy Production from Geothermal Resources*, Report of the Subcommittee on Water and Power Resources, Committee on Interior and Insular Affairs, Washington, D.C.), U.S. Government Printing Office (22-211), December 1973.
(L) Statement of Dr. J. Eggars, Jr., Assistant Director for Research Applications, National Science Foundation: Before the Committee on Interior and Insular Affairs, United States Senate, June 22, 1972.

(L) "Geothermal Electricity Production," *Science* (April 19, 1974), p. 371.

(M) "Environmental Impact of a Geothermal Power Plant," *Science* (March 7, 1975), p. 795.

(L) "Geothermal Energy—the Power under Our Feet," *Science Digest* (July 1974), p. 10.

(L) "Geothermal Power—Tap the Earth's Heat," *Electric Light and Power* (September 1972), p. 56.

(M) "The Convective Earth," *Technology Review* (December 1972), p. 24.

(L) "Geothermal—Earth's Primordial Energy," *Technology Review* (October–November 1971), p. 42.

(L) "Geothermal Energy: An Emerging Major Resource," *Science* (September 15, 1972), p. 978.

(L) "Geothermal Power: Can It Help Solve the Energy Crisis?," *Machine Design* (May 3, 1973), p. 30.

(L) "The Geology and Geophysics of Geothermal Energy," *Technology Review* (March–April 1975), p. 46.

SOLAR POWER

(L) "Solar Energy Now," *Sierra Club Magazine* (May 1974), p. 5.

(L) *Solar Energy*, NASA Release No. 74-47.

(L) "Will Solar Cells Shine on Earth?," *Electronics* (May 22, 1972), p. 67.

(L) "Photovoltaic Cells, Direct Conversion of Solar Energy," *Science* (November 17, 1972), p. 732.

(M) "Satellite Power Stations: A New Source of Energy?," *IEEE Spectrum* (March 1973), p. 38.

(L) "An Ocean-Based Solar-to-Hydrogen Energy Conversion Concept," *Sea Technology* (February 1974), p. 21.

(M) "Plumbing the Ocean's Depths: A New Source of Power," *IEEE Spectrum* (October 1973), p. 22.

(M) "Power in the Year 2001," *Mechanical Engineering* (October 1971), p. 21.

(M) "Solar Sea Power," *Physics Today* (January 1973), p. 48.

(M) "Solar Power via Satellite," *Astronautics and Aeronautics* (August 1973), p. 60.

(L) "Will Solar Farming Solve our Power Crisis?," *Popular Mechanics* (July 1972), p. 90.

(L) Statement of Dr. Alfred J. Eggers, Jr., Assistant Director for Research Applications, National Science Foundations, before the Committee on Interior and Insular Affairs, U.S. Senate, June 7, 1972 (Government Printing Office).

(H) "Solar Energy by Photosynthesis," *Science* (April 19, 1974), p. 375.

(M) "Solar Energy Utilization by Physical Methods," *Science* (April 19, 1974), p. 382.

(L) "Harnessing Solar Energy: The Potential," *Astronautics and Aeronautics* (August 1973), p. 30.

(M) "Getting Bulk Power from Solar Energy," *Electric Light and Power* (December 1974), p. E/G 10.

(L) "Plentiful Energy from the Sun," *Popular Science* (December 1972), p. 87.

(L) "Solar Energy: The Largest Resource," *Science* (September 22, 1972), p. 1088.

(M) "Solar Energy: Its Time Is Near," *Technology Review* (December 1973), p. 31.

(M) "Energy Crisis Spurs Development of Photovoltaic Power Sources," *Electronics* (April 4, 1974), p. 99.

(L) "The Sun Belongs to Everyone," *Financial World* (August 1, 1973), p. 14.

(L) "Solar Energy in Earth Processes," *Technology Review* (March–April 1975), p. 35.

(H) *Solar Cells, Outlook for Improved Efficiency.* National Academy of Sciences, 1972.

(M) *Solar Energy as a National Energy Resource,* NSF/RS/N-73-001, Washington, D.C., Government Printing Office, December 1972.

FUSION POWER

(M) "Laser-Induced Thermonuclear Fusion," *Physics Today* (August 1973), p. 46.

(H) "The Promise of Controlled Fusion," *IEEE Spectrum* (November 1971), p. 24.

(L) "Fusion by Laser," *Scientific American* (June 1971), p. 21.

(L) "Lasers Blast a Shortcut to the Ultimate Energy Solution," *Fortune* (May 1974), p. 221.

(L) "Lasers and Fusion," *Industrial Research* (November 1971), p. 46.

(L) "The Ultimate Power Source?" *Electrical World* (August 1974), p. 40.

(L) "Fusion Power, Types, Status, Outlook," *Power Engineering* (March 1974), p. 46.

(L) "Take a Look at the First Fusion Plant," *Electric Light and Power* (November 1974), p. E/G 12.

(L) "Fusion—Can We Simulate the Sun?," *Electric Light and Power* (September 1972), p. 53.

(L) "Laser Studied as Nuclear Power Trigger," *Aviation Week and Space Technology* (March 25, 1974), p. 46.

(L) "Fusion Reactors as Future Energy Sources," *Science* (November 1, 1974), p. 397.

(L) "Harnessing H-Bomb for Energy," *U. S. News and World Report* (February 17, 1975), p. 69.

(L) "A Future ICE (Thermonuclear, That Is)," *IEEE Spectrum* (January 1973), p. 60.

(M) "The Prospects of Fusion Power," *Scientific American* (February 1971), p. 50.

(M) "Power from Laser-Initiated Nuclear Fusion," *Astronautics and Aeronautics* (August 1973), p. 44.

(M) *Fusion Power* Washington, D.C., Government Printing Office, February, 1973.

Index

333.7 DIC   36432010221729

The electric wishing well :

333.7   DiCerto, Joseph J.
DIC
c.2   The electric wishing well

CHS - U. D.
LIBRARY

12 '93			
Sallman			
GAYLORD			PRINTED IN U.S.A.